Experiments in Basic Circuits
Theory and Application

Third Edition

To Accompany FLOYD, PRINCIPLES OF ELECTRIC CIRCUITS

Fifth Edition

David Buchla
Yuba College

Prentice Hall
Upper Saddle River, New Jersey
Columbus, Ohio

Cover photo: © Superstock
Editor: Linda Ludewig
Developmental Editor: Carol Hinklin Robison
Production Editors: Rex Davidson/Ben Shriver
Cover Designer: Brian Deep
Production Manager: Patricia A. Tonneman
Marketing Manager: Debbie Yarnell

This book was set in Times Roman by The Clarinda Company and was printed and bound by Courier/Kendallville, Inc. The cover was printed by Phoenix Color Corp.

Printed in the United States of America

10 9 8 7 6 5 4 3 2 1

ISBN 0-13-236050-0

Prentice-Hall International (UK) Limited, *London*
Prentice-Hall of Australia Pty. Limited, *Sydney*
Prentice-Hall Canada Inc., *Toronto*
Prentice-Hall Hispanoamericana, S.A., *Mexico*
Prentice-Hall of India Private Limited, *New Delhi*
Prentice-Hall of Japan, Inc., *Tokyo*
Simon & Schuster Asia Pte. Ltd., *Singapore*
Editora Prentice-Hall do Brasil, Ltda., *Rio de Janeiro*

Contents

Preface

The experiments in this Third Edition of the lab manual are designed to be used with *Principles of Electric Circuits,* Fifth Edition, by Thomas Floyd. All experiments have been reviewed and modified as necessary with an eye for improving the directions, readability, and coordination with Floyd's new edition. Two new features have been added to this edition of the lab manual: example PSpice problems that are related to the experiment, and a detailed summary of Floyd's text that can serve as a study guide, review, or as a planning guide for students and instructors. The chapter summaries are given in Appendix A. The example PSpice problems can be used in conjunction with the text or used as an optional exercise in the lab. Experiments that have a PSpice example are indicated with a logo. These files have all been tested on MicroSim PSpice evaluation version. Complimentary PSpice software is also available to instructors from MicroSim.*

The experiments are designed to be completed in one lab period with additional activities suggested by the sections entitled For Further Investigation and Application Problem. I find that many students want to know how the basic laws and passive circuits they learn in a basic electronics course are applied in more complicated active circuits. Accordingly, I have included applications from a number of areas in electronics for each basic experiment. The applications are intended to show the road ahead—with concrete examples requiring application of basic laws to situations they are likely to encounter in the future. It is my hope that these application problems will make learning the basic principles more rewarding by giving students a clear idea of why these principles are important.

Each experiment contains the following parts:

Objectives: Statement of what the student should be able to do after completing the experiment.
Reading: Reading assignments, which are referenced to Floyd's text.
Materials Needed: A list of the components and small items required but not including the equipment found at a typical lab station. Every attempt has been made to specify materials that are common and readily available in typical electronics laboratories. Materials needed for the For Further Investigation and for the Application Problem sections are listed as optional.
Summary of Theory: The Summary of Theory is intended to reinforce the important concepts in Floyd's text with a review of the main points prior to the laboratory experience. In some cases, specific practical information needed in the experiment is presented.
Procedure: This section contains a relatively structured set of steps for performing the experiment. Laboratory techniques, such as operation of the oscilloscope, are given in detail.
For Further Investigation: This section contains specific suggestions for additional related laboratory work. A number of these lend themselves to a formal laboratory report, or they can be used as an enhancement.
Application Problem: This is additional experimental work illustrating how the laws or circuits developed in the experiment are applied to "real world" situations. The circuits are generally simplified and presented with sufficient theory to illustrate the basic experiment. The procedure is not as specific as in the main part of the experiment, allowing the student to investigate the circuit in his or her own way. This section can be assigned as an enchantment or taken up as part of a class review of the experiment.

---

*Contact Product Marketing Dept., MicroSim Corporation, 20 Fairbanks, Irvine, CA 92718

Report Section: The experimental report is formatted in a manner similar to standard laboratory notebooks. To encourage student awareness of the objectives for the experiment, students should complete the blank Objectives section. The Procedure section is primarily intended for the student to either reference the lab book procedure or to document any substitutions, deletions, or changes that were made. Necessary blank tables and graphs are in the Data section. The Results and Conclusion section is left for the student to complete a written summary of key findings for the experiment. Room is left in the report section for summarizing the For Further Investigation and Application Problem results also.
Evaluation and Review Questions: This section contains five or six questions which require the student to draw conclusions from the laboratory work and check his or her understanding of the concepts. Troubleshooting questions are frequently presented.

Each lab station should contain a dual variable regulated power supply, a function generator, a multimeter, and a dual-channel oscilloscope. It is useful if the laboratory is equipped to measure capacitors and inductors. In addition, a meter calibrator and a commercial Wheatstone bridge are useful but not required. A list of all required components is given in Appendix B.

All of the experiments in this manual have been tested in classrooms with most experiments having been performed by literally thousands of students. While this manual is specifically designed to follow the sequence of Floyd's *Principles of Electric Circuits* text, it can be used with other texts by ignoring the references.

I appreciate the efforts of my colleagues at Yuba College, Wayne Cromwell and Bill Frandrup, for their assistance and ideas. I would also like to express my appreciation to the following reviewers who have reviewed previous editions of this manual: Harvey Laabs, North Dakota State College of Science; Arnold Kroeger, Hillsborough Community College; and Ulrich E. Zeisler, Salt Lake Community College. As always, it is a pleasure to work with the staff at Prentice Hall Publishing Company. I thank Linda Ludewig, Carol Robison, and Rex Davidson and the entire production team for their commitment to quality. Finally, I express my appreciation to my wife, Lorraine, for her continued support and for not disturbing my piled higher and deeper (PhD) filing system.

David Buchla

Introduction to the Student

The purpose of experimental work is to help you gain a better understanding of the principles of electronics and to give you experience with instruments and methods used by technicians and electronic engineers. You should begin each experiment with a clear idea of the purpose of the experiment and the theory behind the experiment. Each experiment will require you to use electronic instruments to measure various quantities. The measured data will be recorded and you will need to interpret the measurements and draw conclusions about your work. The ability to measure, interpret, and communicate results is basic to electronic work.

Each experiment in this experiment manual contains a report section with headings similar to those used by technicians who must keep a laboratory notebook. Before the experiment, you should read it over and know what you are trying to accomplish in the experiment. Write, in your own words, the experimental objectives into the laboratory report under the section titled *Objectives*. The report has a short section entitled *Procedure*. If you followed the procedure given in the directions, you can make a notation such as "See procedure in lab book." If you needed to change or omit a procedure, then a note as to what change was necessary should be made. For example, if you needed to substitute a different value component, this should be noted. Data tables and plots are provided as needed. You will need to interpret and discuss the results in the section titled *Results and Conclusion* and in the *Evaluation and Review Questions*. Conclusions are your interpretation of the data and should include important findings with a brief discussion of problems, or revisions, or suggestions you may have. This is a good place to discuss measurement errors. The conclusion should answer, in some way, the objectives of the experiment. For example, if the purpose of the experiment was to determine the frequency response of a filter, the conclusion should describe the frequency response or contain a reference to an illustration of the response. Your instructor may assign additional work in the section titled *For Further Investigation* or the *Application Problem*. These sections are intended to reinforce the experiment and show you how your study of basic electronics is applied to circuits throughout electronics.

SAFETY IN THE LABORATORY

The experiments in this lab book are designed for low voltages to minimize electric shock hazard; however, one should never assume that electric circuits are safe. A current through the body of a few milliamps can be lethal. In addition, electronic laboratories often contain other hazards such as chemicals and power tools. For your safety, you should review laboratory safety rules before beginning a course in electronics. In particular, you should do the following:

1. Avoid contact with *any* voltage source. Turn off power before working on circuits.
2. Remove watches, jewelry, rings, and so forth before working on circuits—even those circuits with low voltages, as burns can occur.
3. Know the location of the emergency power-off switch.
4. Never work alone in the laboratory.
5. Ensure that line cords are in good condition and grounding pins are not missing or bent. Do not defeat the three-wire ground system in order to make "floating" measurements.

6. Keep a neat work area and handle tools properly. Wear safety goggles or gloves when required.
7. Report any unsafe condition to your instructor.
8. Be aware of and follow laboratory rules.

The Laboratory Notebook

1. PURPOSE OF A LABORATORY NOTEBOOK

A laboratory notebook forms a chronological record of laboratory work in such a manner that it could be reconstructed if necessary. The notebook is a bound and numbered daily record of laboratory work. Data are recorded as they are observed. Each page is dated as it is done and the signature of the person doing the work is added to make the work official, since laboratory notebooks may be the basis of patent applications or have other legal purposes. No pages are left blank and no pages may be removed.

2. GENERAL INFORMATION

The format of laboratory notebooks may vary; however, certain requirements are basic to all laboratory notebooks. The name of the experimenter, date, and purpose of the experiment are entered at the top of each page. All test equipment should be identified by function, manufacturer, and serial number to facilitate reconstruction of the experiment. Test equipment may be identified on a block diagram or circuit drawing rather than an equipment list. References to books, articles or other sources that were used in preparing for the experiment are noted if they were used. A *brief* description of the procedure is necessary. The procedure is not a restatement of the instructions in the experiment book but rather is a concise statement about what *was* done in the experiment.

3. RECORDING OF DATA

Data taken in an experiment should be directly recorded in tabular form in the notebook. Raw (not processed) data should be recorded. Data should not be transcribed from scratch paper. When calculations have been applied to data, a sample calculation should be included to indicate clearly what process has been applied to the raw data. If an error is made, a single line should be drawn through the error with a short explanation.

When a measurement contains approximate numbers, the digits known to be correct are called significant digits. The number of significant digits for a given measurement depends on the quality of the instruments used. As a rule of thumb, for laboratory work in basic electronics, three significant digits should be retained. Significant digits are discussed further in the *Summary of Theory* for Experiment 1.

4. GRAPHS

A graph is a visual tool that can quickly convey to the reader the relationship between variables. The eye can discern trends in magnitude or slope more easily than from tabular data. Graphs are constructed with the independent variable plotted along the horizontal axis (called the *abscissa*) and the dependent variable plotted along the vertical axis (called the *ordinate*). A smooth curve can be drawn showing the trend of the data. It is not necessary to connect the data points (except in calibration curves). For data in which one of the variables is related to the other by a power, logarithmic (log) scales on one or both axes may show the relationship of data. Log scales can show data over a large range of values that will not fit on ordinary graph paper.

When you have determined the type of scale that best shows the data, select numbers for the scale that are easily read. Do not use the data for the scale; rather, choose numbers that allow the largest data point to fit on the graph. Scales should generally start at zero unless limitations in the data preclude it. Data points on the graph should be shown with a dot in the center of a symbol such as a circle or triangle. The graph should be labeled in a manner that is self-explanatory. A figure number should be used as a reference in the written explanation.

5. SCHEMATICS AND BLOCK DIAGRAMS

Schematics, block diagrams, waveform drawings, and other illustrations are important tools for depicting the facts of an experiment. Experiments with circuits need at least a schematic drawing showing the setup and may benefit from other illustrations, depending on your purpose. Usually, simpler drawings are best; however, sufficient detail needs to be shown to enable the reader to reconstruct the circuit if necessary. Adequate information should be contained in an illustration to make its purpose clear to the reader.

6. RESULTS AND CONCLUSION

The section that discusses the results and conclusion is the most important part of your laboratory notebook, because it contains the key findings of your experiment. Each experiment should contain a conclusion, which is a *specific* statement about the important results you obtained and enables the reader to understand the results properly. Be careful about sweeping generalizations not warranted by the experiment. Before writing a conclusion, it is useful to review the purpose of the experiment. A good conclusion "answers" the purpose of the experiment. For example, if the purpose of an experiment is to determine the frequency response of a filter, the conclusion should describe the frequency response or contain a reference to an illustration of the response. In addition, the conclusion should contain an explanation of difficulties, unusual results, revisions or any suggestions you may have for improving the circuit.

7. SUGGESTED FORMAT

From the foregoing discussion, the following format is suggested. This format may be modified as circumstances dictate:

1. *Title and date.*
2. *Purpose.* Give a statement of what you intend to determine as a result of the investigation.
3. *Equipment and materials.* Include equipment model and serial numbers, which can allow retracing if a defective or uncalibrated piece of equipment was used.
4. *Procedure.* Include a description of what you did and what measurements you made. A reference to the schematic drawing should be included.
5. *Data.* Tabulate raw (unprocessed) data; data may be presented in graphical form.
6. *Sample calculations.* If you have a number of calculations, then give a sample calculation showing the formulas that you applied to the raw data to transform it to processed data. This section may be omitted if calculations are clear from the procedure or they are discussed in the results.
7. *Results and conclusion.* This section is the place to discuss your results, including experimental errors. This section should contain key information about the results and your interpretation of the significance of these results.

Each page of the laboratory notebook should be signed and dated as previously discussed.

The Technical Report

1. EFFECTIVE WRITING

The purpose of technical reports is to communicate technical information in a way that is easy for the reader to understand. Effective writing requires that you know your reader's background. You must be able to put yourself in the reader's place and anticipate what information you must convey to have the reader understand what you are trying to say. When you are writing experimental results for a person working in your field, such as an engineer, your writing style may contain words or ideas that are unfamiliar to a layperson. If your report is intended for persons outside your field, you will need to provide background information.

2. WORDS AND SENTENCES

You need to choose words that have clear meaning to a general audience or define every term that does not have a well-established meaning. Keep sentences short and to the point. Short sentences are easiest for the reader to comprehend. Avoid stringing a series of adjectives or modifiers together. For example, this figure caption contains an unclear string of modifiers:

Operational amplifier constant-current source schematic

The word schematic has been modified five times. By changing the order and adding natural connectors such as *of, using,* and *an,* the meaning can be clarified:

Schematic of a constant-current source using an operational amplifier

3. PARAGRAPHS

Paragraphs need to contain a unit of thought. Excessively long paragraphs suffer from the same weakness that afflict overly long sentences. The reader is asked to digest too much material at once, causing comprehension to diminish. Paragraphs should organize your thoughts in a logical format. Look for natural breaks in your ideas. Each paragraph should have one central idea and contribute to the development of the entire report.

Good organization is the key to a well-written report. Outlining in advance will help organize your ideas. The use of headings and subheadings for paragraphs or sections can help steer the reader through the report. Subheadings also prepare the reader for what is ahead and make the report easier to understand.

4. FIGURES AND TABLES

Figures and tables are effective ways to present information. Figures should be kept simple and to the point. Often a graph can make clear the relationship of data. Comparisons of different data drawn on the same graph make the results more obvious to the reader. Figures should be labeled with a figure number and a brief label. Don't forget to label both axes of graphs.

Data tables are useful for presenting data. Usually data presented in a graph or figure should not also be included in a data table. Data tables should be labeled with a table number and short title. The data table should contain enough information that its meaning is clear to the reader without having to refer to the text. If the purpose of the table is to compare information, then form the data in columns rather than rows. Information in columns is easier for people to compare. Table footnotes are a useful method of clarifying some point about the data. Footnotes should appear at the bottom of the table, with a key to where the footnote applies.

Data should appear throughout your report in consistent units of measurement. Most sciences use the metric system; however, the English (or customary) system is still sometimes used. The metric system uses derived units that are cgs (centimeter-gram-second) or mks (meter-kilogram-second). It is best to use consistent metric units or to include a conversion chart. Whatever units are used should be used throughout your report.

Tabular data should be shown with a number of significant digits consistent with the precision of the measurement. Significant digits are discussed in the *Summary of Theory* for Experiment 1.

Reporting numbers using powers of 10 can be a sticky point with reference to tables. Table 1 shows four methods of abbreviating numbers in tabular form. The first column is unambiguous, since the number is presented in conventional form. This requires more space than if the information is presented in scientific notation. In column 2, the same data are shown with a metric prefix used for the unit. In column 3, the power of 10 is shown. Each of the first three columns shows the measurement unit and is not subject to misinterpretation. Column 4, on the other hand, is wrong. In this case the author is trying to tell us what operation was performed on the numbers to obtain the values in the column. This is incorrect, because the column heading should contain the unit of measurement for the numbers in the column.

Table 1. Reporting numbers in tabular data

Column 1	**Column 2**	**Column 3**	**Column 4**
Resistance ohms	Resistance kΩ	Resistance $\times 10^3$ ohms	Resistance ohms $\times 10^{-3}$
470,000	470	470	470
8,200	8.2	8.2	8.2
1,200,000	1,200	1,200	1,200
330	0.33	0.33	0.33
	Correct		**Wrong**

5. SUGGESTED FORMAT

1. *Title*. A good title needs to convey the substance of your report by using key words that provide the reader with enough information to decide if the report should be investigated further.
2. *Contents*. Key headings throughout the report are listed with page numbers.
3. *Abstract*. The abstract is a *brief* summary of the work with principal facts and results stated in concentrated form. It is a key factor in helping a reader to determine if he or she should read further.
4. *Introduction*. The introduction orients a reader. It should briefly state what you did and give the reader a sense of the purpose of the report. It may tell the reader what to expect and briefly describe the report's organization.

5. *Body of the report*. The report can be made clearer to the reader if you use headings and subheadings to mark major divisions through your report. The headings and subheadings can be generated from the outline of your report. Figures and tables should be labeled and referenced in the body of the report.
6. *Conclusion*. The conclusion summarizes important points or results. It may refer to figures or tables previously discussed in the body of the report to add emphasis to significant points. In some cases, the primary reasons for the report are contained within the body and a conclusion is deemed to be unnecessary.
7. *References*. References are cited to enable the reader to find information used in developing your report or work that supports your report. The reference should include names of all authors, in the order shown in the original document. Use quotation marks around portions of a complete document such as a journal article or a chapter of a book. Books, journals, or other complete documents should be underlined. Finally, list the publisher, city, date, and page numbers.

Reference Guide to Laboratory Instruments

This section is provided to help familiarize you with basic laboratory instruments and may be used as a reference as the instruments are introduced. It is not possible to cover all possible variations between instruments, so only general features common to a class of instruments are described. Consult the operator's manual for detailed descriptions and safe operating practice for the particular instruments in your laboratory.

THE POWER SUPPLY

Most electronic circuits require a source of regulated direct current (dc) to operate properly. A direct current–regulated power supply is a circuit that provides the energy to allow electronic circuits to function. It does this by transforming a source of input electrical power (generally ac) into dc. Most regulated supplies are designed to maintain a fixed voltage that will stay within certain limits of voltage for normal operation. Voltage adjustment and current limits depend on the particular supply.

The power supply must provide the proper level of dc voltage for a given circuit. Some integrated circuits, for example, can function properly only if the voltage is within a very narrow range. You will normally have to set the voltage to the proper level *before* you connect a power supply to the test circuit. The power supply at your bench may have more than one output and normally will have a built-in meter to help you set the voltage. Some power supplies have meters that monitor both voltage and current. The meter may have multiple or complex scales, depending on the available outputs from the supply. These points are discussed in more detail in the first experiment.

It is important that the user make good connections to the power supply output terminals with wire that is sufficient to carry the load current if the output were accidentally shorted together. Clip-leads are not recommended, since they can produce measurement error due to high contact resistance. In situations where several circuits are operated from the same supply, the best policy is to operate each circuit with an independent set of leads.

THE MULTIMETER

The digital multimeter (DMM) and analog volt-ohm-milliammeter (VOM) are multipurpose measuring instruments that combine the characteristics of a dc and ac voltmeter, dc and ac ammeter, and an ohmmeter in one instrument. The DMM indicates the measured quantity as a digital number, avoiding the necessity to interpret the scales, as is required on analog instruments. Although in most labs the DMM has replaced the VOM as the instrument of choice, there are several advantages to the VOM. It is less susceptible to interference and generally has a much higher frequency response.

Because the multimeter is a multipurpose instrument, it is necessary to determine which controls select the desired function. In addition, current measurements (and often high-range voltage measurements) usually require a separate set of lead connections to the meter. After you have selected the function, you need to select the appropriate range to make the measurement. It is important to select the function and range *before* connecting the meter to the circuit you are testing. DMMs can be autoranging, meaning that the instrument automatically selects the correct scale and sets the decimal place, or they can be manual ranging, meaning that the user must select the correct scale. For manual ranging instruments, when the approximate voltage or current is not known, always begin a measurement on the highest possible range to avoid instrument overload and possible damage. Change to a lower range as necessary to increase the precision. The life of range switches will be lengthened if you change ranges only with the probes disconnected from the circuit. On analog instruments the range selected should give a reading in the upper portion of the scale.

The voltmeter function of a DMM can measure either ac or dc volts. The dc voltage function is useful to measure the dc voltage *difference* between two points. If the meter's red lead is touching a more positive point than the meter's black lead, the reading on the meter will be positive; if the black lead is on the more positive point, the reading will be negative. Analog meters *must* be connected with the correct polarity, or the pointer will attempt to move backward, possibly damaging the movement.

The ac voltage function is designed to measure low-frequency sinusoidal waveforms. The reading on a meter is calibrated to read the rms (root-mean-square) value of a sinusoidal waveform. Frequency is the number of cycles per second, measured in hertz, for a waveform. All DMMs and VOMs are limited to some specified frequency range. The meter reading will be inaccurate if you attempt to measure waveforms outside the meter's specified frequency range. A typical DMM is not accurate on the ac scale below about 45 Hz or above about 1 kHz, although this range can be considerably better in some cases. A VOM can measure ac waveforms over a much larger range—typically up to 100 kHz.

The ohms function (used for resistance measurements) is used only in circuits that are *not* powered. An ohmmeter works by inserting a small test voltage into a circuit and measuring the resulting current flow. Consequently, if any voltage is present, the reading will be in error. The meter will show the resistance of all possible paths between the probes. If you want to know the resistance of a single component, it is necessary to isolate that component from the remainder of the circuit by disconnecting one end. In addition, body resistance can affect the reading if you are holding the conducting portion of both probes in your fingers. This procedure should be avoided, particularly with high resistances.

THE FUNCTION GENERATOR

The basic function generator is used to produce sine, square, and triangle waveforms and may also have a pulse output for testing digital logic circuits. Function generators normally have controls that allow you to select the type of waveform and other controls to adjust the amplitude and dc level. The peak-to-peak voltage is adjusted by the AMPLITUDE control. The dc level is adjusted by a control labeled DC OFFSET; this enables you to add or subtract a dc component to the waveform. These controls are generally not calibrated, so amplitude and dc level settings need to be verified with an oscilloscope or multimeter.

The frequency is selected with a combination of a range switch and vernier control. The range is selected by a decade frequency switch or pushbuttons that enable you to select the frequency in decade increments (factors of 10) up to about 1 MHz. The vernier control is usually a multiplier dial for adjusting the precise frequency needed.

The output level of a function generator will drop from its open-circuit voltage when it is connected to a circuit. Depending on the conditions, you generally will need to readjust the amplitude level of the generator after it is connected to the circuit. This is because there is effectively an internal generator resistance called the output resistance that will affect the circuit under test. You can measure this resistance indirectly as described in Experiment 12, *For Further Investigation*.

Better-quality instruments include features such as trigger outputs, modulation, increased frequency ranges, fixed attenuators on the output, and so forth. Details of the particular features of your function generator and the controls can be found in the operator's manual supplied with the instrument.

THE OSCILLOSCOPE

An oscilloscope is the most versatile general-purpose measuring instrument, letting you see a graph of the voltage as a function of time in a circuit. Many circuits have specific timing requirements or phase relationships that can be readily measured with a two-channel oscilloscope. The voltage to be measured is converted into a visible display by a cathode-ray tube (CRT), a vacuum device similar to a television picture tube.

The oscilloscope contains four functional blocks, as illustrated in Figure 17–1. The input signal is connected to the *vertical* section, which can be set to attenuate or amplify the input signal to provide the proper voltage level to the vertical deflection plates of the CRT. The *trigger* section samples the input waveform and sends a synchronizing trigger signal at the proper time to the horizontal section. The trigger occurs at the same relative time thus superimposing each succeeding trace on the previous trace. This action causes the signal to appear to stop, allowing you to examine the signal. The *horizontal* section contains the time-base (or *sweep*) generator, which produces a linear ramp, or "sawtooth," waveform that controls the rate the beam moves across the screen. The horizontal position of the beam is proportional to the time that elapsed from the start of the sweep, allowing the horizontal axis to be calibrated in units of time. For this reason, the horizontal section is often called the *time base*. The output of the horizontal section is applied to the horizontal deflection plates of the CRT. Finally, the *display* section contains the CRT and beam controls. It enables the user to obtain a sharp presentation with the proper intensity. The display section frequently contains other features. The user can sometimes lose the displayed waveform by accidentally positioning it offscreen. One common feature in the display section is a beam-finder button, which enables the user quickly to locate the position of the trace. Controls for each of the functional blocks are usually grouped together. Frequently, there are color clues to help you identify groups of controls. Details of these controls are explained in the operator's manual for the oscilloscope; however, a brief description of frequently used controls is given in the following paragraphs.

Display Controls

The display system contains controls for adjusting the electron beam, including FOCUS and INTENSITY controls. FOCUS and INTENSITY are adjusted for a comfortable viewing level with a sharp focus. The display section may also contain the BEAM FINDER, a control used in combination with the horizontal and vertical POSITION controls to bring the trace on the screen. Another control over the beam intensity is the *z*-axis input. A control voltage on the *z*-axis input can be used to turn the beam on or off or adjust its brightness. Some oscilloscopes also include the TRACE ROTATION control in the display section. TRACE ROTATION is used to align the sweep with a horizontal graticule line. This control is usually adjusted with a screwdriver to avoid accidental adjustment.

Vertical Controls

The vertical controls include the VOLTS/DIV (vertical sensitivity) control and its vernier, the input COUPLING switch, and the vertical POSITION control. Dual-trace and multiple-channel oscilloscopes will have a duplicate set of these controls for each channel and various switches for selecting channels or other vertical operating modes. The vertical inputs is connected through a selectable attenuator to a high input impedance dc amplifier. The VOLTS/DIV control selects a combination of attenuation/gain to determine the vertical sensitivity of the scope. For example, a low-level signal will need more gain and less attenuation than a higher level signal. The vertical sensitivity is adjusted in fixed VOLTS/DIV increments to allow the user to make calibrated voltage measurements. In addition, a concentric vernier control is usually provided to allow a continuous range of sensitivity. This knob must be in the detent (calibrated) position to make voltage measurements. The detent position can be felt by the user as the knob is turned because the knob tends to "lock" in the detent position. Some oscilloscopes have a warning light or message when the vernier is not in its detent position.

The input coupling switch is a multiple-position switch than can be set for AC, GND, or DC and sometimes includes a 50 Ω position. The GND position of the switch internally disconnects the signal from the scope and grounds the input amplifier. This position is useful if you want to set a ground reference level on the screen for measuring the dc component of a waveform. The AC and DC positions are high-impedance inputs—typically 1 MΩ shunted by 15 pF of capacitance. High-impedance inputs are useful for general probing at frequencies below about 1 MHz. At higher frequencies, the shunt capacitance can load the signal source excessively, causing measurement error. Attenuating divider probes are good for high-frequency probing because they have very high impedance (typically 10 MΩ) with very low shunt capacitance (as low as 2.5 pF).

The AC position of the coupling switch inserts a series capacitor before the input attenuator, causing dc components of the signal to be blocked. This position is useful if you want to measure a small ac signal riding on top of a large dc signal—power supply ripple, for example. The DC position is used when you want to view *both* the AC and DC components of a signal. This position is best when viewing digital signals, because the input *RC* circuit forms a differentiating network. The AC position can distort the digital waveform because of this differentiating circuit. The 50 Ω position places an accurate 50 Ω load to ground. This position provides the proper termination for probing in 50 Ω systems and reduces the effect of a variable load, which can occur in high-impedance termination. The effect of source loading *must* be taken into account when using a 50 Ω input. It is important not to overload the 50 Ω input, because the resistor is normally rated for only 2 W—implying a maximum of 10 V_{rms} of signal—can be applied to the input.

The vertical POSITION control varies the dc voltage on the vertical deflection plates, allowing you to position the trace anywhere on the screen. Each channel has its own vertical POSITION control, enabling you to separate the two channels on the screen. You can use vertical POSITION when the coupling switch is in the GND position to set an arbitrary level on the screen as ground reference.

Some oscilloscopes also have a vertical magnifier. The magnifier increases the scope sensitivity by providing more gain to the input signal but at the expense of the overall bandwidth. The vertical magnifier is useful for measuring very low level signals that are not bandwidth limited.

Horizontal Controls

The horizontal controls include the SEC/DIV control and its vernier, the horizontal magnifier, and the horizontal POSITION control. In addition, the horizontal section may include delayed sweep controls. The SEC/DIV control sets the sweep speed, which controls how fast the electron beam is moved across the screen. The control has a number of calibrated positions divided into steps of 1-2-5 multiples, which allow you to set the exact time interval at which you view the input signal. For example, if the graticule has 10 horizontal divisions and the SEC/DIV control is set to 1.0 ms/div, then the screen will show a total time of 10 ms. The SEC/DIV control usually has a concentric vernier control that allows you to adjust the sweep speed continuously between the calibrated steps. This control must be in the detent position in order to make calibrated time measurements. Many scopes are also equipped with a horizontal magnifier that affects the time base. The magnifier increases the sweep time by the magnification factor, giving you increased resolution of signal details. Any portion of the original sweep can be viewed using the horizontal POSITION control in conjunction with the magnifier. This control actually speeds the sweep time by the magnification factor and therefore affects the calibration of the time base set on the SEC/DIV control. For example, if you are using a 10X magnifier, the SEC/DIV dial setting must be *divided* by 10.

Trigger Controls

The trigger section is the source of most difficulties when learning to operate an oscilloscope. These controls determine the proper time for the sweep to begin in order to produce a stable display. The trigger controls include the MODE switch, SOURCE switch, trigger LEVEL, SLOPE, COUPLING, and variable HOLDOFF controls. In addition, the trigger section includes a connector for applying an EXTernal trigger to start the sweep. Trigger controls may include HIGH or LOW FREQUENCY REJECT switches and BANDWIDTH LIMITING.

The MODE switch is a multiple-position switch that selects either AUTO or NORMAL (sometimes called TRIGGERED) and may have other positions, such as TV or SINGLE sweep. In the AUTO position, the trigger generator selects an internal oscillator that will trigger the sweep generator as long as no other trigger is available. This mode ensures that a sweep will occur even in the absence of a signal, because the trigger circuits will "free run" in this mode. This allows you to obtain a baseline for adjusting ground reference level or for adjusting the display controls. In the NORMAL or TRIGGERED mode, a trigger is generated from one of three sources selected by the SOURCE switch—the INTERNAL signal, an EXTERNAL trigger source, or the ac LINE. If you are using the internal signal to obtain a trigger, the normal mode will provide a trigger only if a signal is present and other trigger conditions (level, slope) are met. This mode is more versatile than AUTO as it can provide stable triggering for very low to very high frequency signals. The TV position is used for synchronizing either television fields or lines and SINGLE is used primarily for photographing the display.

The trigger LEVEL and SLOPE controls are used to select a specific point on either the rising or falling edge of the input signal for generating a trigger. The trigger SLOPE control determines which edge will generate a trigger, whereas the LEVEL control allows the user to determine the voltage level on the input signal that will start the sweep circuits.

The SOURCE switch selects the trigger source—either the INTERNAL signal, an EXTERNAL trigger source, or the ac LINE. In the INTERNAL position, a sample of the signal being viewed is used to start the sweep. Some multiple-channel scopes allow you to choose the triggering channel or use some combination of channels for triggering. In the EXTERNAL position, a time-related external signal is used for triggering. Trigger attenuation (divide by

10) is provided on many oscilloscopes. Large external triggers can be attenuated in order to extend the range of the trigger level control. The external trigger can be coupled with either AC or DC COUPLING. The trigger signal can be coupled with AC COUPLING if the trigger signal is riding on a dc voltage. DC COUPLING is used if the triggers occur at a frequency of less than about 20 Hz. The LINE position causes the trigger to be derived from the ac power source. This synchronizes the sweep with signals that are related to the power line frequency.

The variable HOLDOFF control allows you to exclude otherwise valid triggers until the holdoff time has elapsed. For some signals, particularly complex waveforms or digital pulse trains, obtaining a stable trigger can be a problem. This can occur when one or more valid trigger points occurs before the signal repetition time. If every event that the trigger circuits qualified as a trigger were allowed to start a sweep, the display could appear to be unsynchronized. By adjusting the variable HOLDOFF control, the trigger point can be made to coincide with the signal-repetition point.

Occasionally, a trigger signal contains interfering high- or low-frequency components. When this happens, the trace may flutter back and forth on the screen or produce the appearance of unsynchronized sweeps. A high-frequency component can be removed using HIGH FREQUENCY REJECT. Sometimes it is necessary to observe a high-frequency signal in the presence of a low-frequency signal such as might occur in measuring low-level transducer signal. In this case, LOW FREQUENCY REJECT can remove the low-level interference from the trigger signal.

Dual-Channel Oscilloscopes

Most oscilloscopes have two (or in some cases more) separate channels and controls to view more than one signal at a time. There are two types of dual-channel oscilloscope—dual beam and dual trace. A dual-beam oscilloscope has two independent beams in the CRT and independent vertical deflection systems, allowing both signals to be viewed at the same time. A dual-trace oscilloscope has only one beam and one deflection system; it uses electronic switching to show the two signals. Dual-beam oscilloscopes are generally restricted to high-performance research instruments and are much more expensive than dual-trace oscilloscopes.

A dual-trace oscilloscope has user controls labeled CHOP or ALTERNATE to switch the beam between the channels so that the signals appear to occur simultaneously. The CHOP mode rapidly switches the beam between the two channels at a fixed high speed rate, so the two channels appear to be displayed at the same time. The ALTERNATE mode first completes the sweep for one of the channels and then displays the other channel on the next (or *alternate*) sweep. When viewing slow signals, the CHOP mode is best, because it reduces the flicker that would otherwise be observed. High-speed signals can usually be observed best in ALTERNATE mode to avoid seeing the chop frequency.

Another feature on most dual-trace oscilloscopes is the ability to show the algebraic sum and difference of the two channels. For most measurements, you should have the vertical sensitivity (VOLTS/DIV) on the same setting for both channels. You can use the algebraic sum if you want to compare the balance on push-pull amplifiers, for example. Each amplifier should have identical out-of-phase signals. When the signals are added, the resulting display should be a straight line, indicating balance. You can use the algebraic difference when you want to measure the waveform across an ungrounded component. The probes are connected across the ungrounded component with probe ground connected to circuit ground. Again, the vertical sensitivity (VOLTS/DIV) setting should be the same for each channel. The display will show the algebraic difference in the two signals. The algebraic difference mode also allows you to cancel any unwanted signal that is equal in amplitude and phase and is common to both channels.

Dual-trace oscilloscopes usually have an X-Y mode, which causes one of the channels to be graphed on the *x*-axis and the other channel to be graphed on the *y*-axis. This is necessary if you want to change the oscilloscope base line to represent a quantity other than time. Applications include viewing a transfer characteristic (output voltage as a function of input voltage), swept frequency measurements, or showing Lissajous figures for phase measurements. Lissajous figures are patterns formed when sinusoidal waves drive both channels and are described in Experiment 25, *For Further Investigation*.

Probes

Signals should always be coupled into the oscilloscope through a probe. A probe is used to pick off a signal and couple it to the oscilloscope. It reduces the loading effect on circuits. Probes also have a short ground lead that should be connected to a nearby circuit ground point to avoid oscillation and power line interference. The ground lead makes a mechanical connection to the test circuit and passes the signal through a flexible, shielded cable to the oscilloscope. The shielding helps protect the signal from external noise pickup. In addition, the probe can extend the oscilloscope's measuring ability by reducing the effects of circuit loading and extending the amplitude capability of the scope. Oscilloscope probes are provided with the instrument as part of the system. The wrong probe—or one that is not adjusted properly—can affect the measurement to the point of making it worthless or misleading.

Begin any session with the oscilloscope by checking the probe compensation on each channel. Adjust the probe for a flat-topped square wave while observing the scope's calibrator output. This is a good signal to check the focus and intensity and verify trace alignment. Check the front-panel controls for the type of measurement you are going to make. Normally, the variable controls (VOLTS/DIV and SEC/DIV) should be in the calibrated (detent) position. The vertical coupling switch is usually placed in the DC position unless the waveform you are interested in has a large dc offset. Trigger holdoff should be in the minimum position unless it is necessary to delay the trigger to obtain a stable sweep.

1 Metric Prefixes and Scientific Notation

OBJECTIVES:

After performing this experiment, you will be able to:

1. Convert standard form numbers to scientific and engineering notation.
2. Measure quantities using a metric prefix.
3. Identify several electronic components.

READING:

Floyd, *Principles of Electric Circuits,* Sections 1–4 through 1–5

MATERIALS NEEDED:

Scientific calculator
Metric ruler

SUMMARY OF THEORY:

Persons working in electronics need to be able to make concise statements about measured quantities. Symbols are used to represent electrical and magnetic quantities as shown in Floyd's text. The basic electrical quantities encompass a very large range of numbers—from the very large to the very small. For example, the frequency of an FM radio station can be over 100 million hertz (Hz) and a capacitor can have a value of 10 billionths of a farad (F). To express very large and very small numbers, scientific (powers of 10) notation and metric prefixes are used. Metric prefixes are based on the decimal system and stand for powers of 10. They are widely used to indicate a multiple or submultiple of a measurement unit.

Scientific notation is a means of writing any quantity as a number between 1 and 10 times a power of 10. The power of 10 is called the *exponent*. It simply shows how many places the decimal place must be shifted to express the number in its standard form. If the exponent is positive, the decimal place must be shifted to the right to write the number in standard form. If the exponent is negative, the decimal place must be shifted to the left. Note that $10^0 = 1$, so multiplying by a power of 10 with an exponent of zero does not change the original number.

Exponents that are a multiple of 3 are much more widely used in electronics work than exponents that are not multiples of 3. Numbers expressed with an exponent that is a multiple of 3 are said to be expressed in *engineering notation*. Engineering notation is particularly useful in electronics work because of its relationship to the most widely used metric prefixes. Some examples of numbers written in standard form, scientific notation, and engineering notation are shown in Table 1–1.

Numbers expressed in engineering notation can be simplified by using metric prefixes to indicate the appropriate power of 10. In addition, prefixes can simplify calculations. You

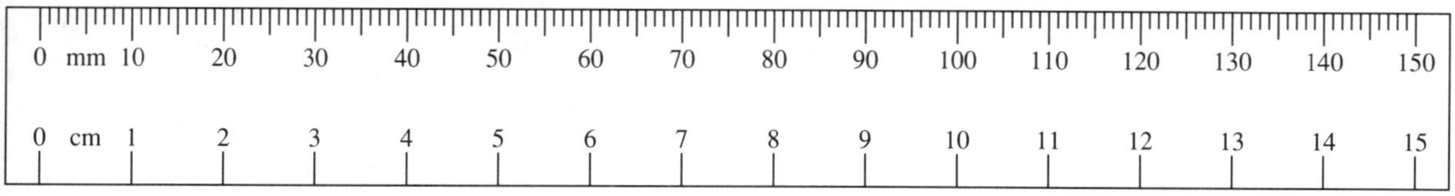

Table 1–1

Standard Form	Scientific Notation	Engineering Notation
12,300.	1.23×10^4	12.3×10^3
123.	1.23×10^2	0.123×10^3
1.23	1.23×10^0	1.23×10^0
0.0123	1.23×10^{-2}	12.3×10^{-3}
0.000 123	1.23×10^{-4}	$123. \times 10^{-6}$

can perform arithmetic operations on the significant figures of a problem and determine the answers prefix from those used in the problem. For example, 4.7 kΩ + 1.5 kΩ = 6.2 kΩ. The common metric prefixes used in electronics and their abbreviations are shown in Table 1–2. The metric prefixes representing engineering notation are shown. Any number can be converted from one prefix to another (or no prefix) using the table. Write the number to be converted on the line with the decimal under the metric prefix that appears with the number.

Table 1–2

Power of 10:	10^9	10^6	10^3	10^0	10^{-3}	10^{-6}	10^{-9}	10^{-12}
Metric symbol:	G	M	k		m	μ	n	p
Metric prefix:	giga	mega	kilo		milli	micro	nano	pico

0 | 000 | 000 | 000 . 000 | 000 | 000 | 000 | 0

Example 1:

Convert 12,300,000 to a number with a metric prefix:

Metric prefix:	giga	mega	kilo	milli	micro	nano	pico

0 | 000 | 000 | 000 . 000 | 000 | 000 | 000 | 0

12 300 000 .

= 12 . 3M

Example 2:

Change 10,000 pF to a number with a μ prefix:

Metric prefix:	giga	mega	kilo	milli	micro	nano	pico

0 | 000 | 000 | 000 . 000 | 000 | 000 | 000 | 0

10 000 . pF

. 0 1 0 μF

The decimal place is then moved directly under any other line and the metric prefix immediately above the line is used. The number can also be read in engineering notation by using the power of 10 shown immediately above the line.

Scientific calculators have the ability to process numbers that are written in exponential form. In addition, scientific calculators can perform trig functions, logarithms, roots, and other math functions. To enter numbers in scientific notation, the base number (called the *mantissa*) is first entered. If the number is negative, the +/− key is pressed. Next the exponent is entered by pressing the EE (or EXP) key, followed by the power of 10. If the exponent is negative, the +/− key is pressed. Arithmetic can be done on the calculator with numbers in scientific notation mixed with numbers in standard form.

SIGNIFICANT DIGITS:

When a measurement contains approximate data, those digits known to be correct are called *significant digits*. Zeros that are used only for locating the decimal place are *not* significant, but those that are part of the measured quantity are significant. When reporting a measured value, the least significant uncertain digit may be retained, but all other uncertain digits should be discarded. It is *not* correct to show either too many or too few digits. For example, it is not valid to retain more than three digits when using a meter that has three digit resolution, nor is it proper to discard valid digits, even if they are zeros. For example, if you set a power supply to the nearest hundredth of a volt, then the recorded voltage should be reported to the hundredth place (3.00 V is correct, but 3 V is incorrect). For laboratory work in this course, you should normally be able to measure and retain three significant digits.

To find the number of significant digits in a given number, ignore the decimal place and count the number of digits from left to right, starting with the first nonzero digit and ending with the last digit to the right. All digits counted are significant except zeros at the right end of the number. A zero on the right end of a number is significant *only* if it is to the right of the decimal place; otherwise it is uncertain. For example, 43.00 contains four significant digits. The whole number 4300 may contain two, three, or four significant digits. In the absence of other information, the significance of the right-hand zeros is uncertain, and these digits cannot be assumed to be significant. To avoid confusion, numbers such as these should be reported using scientific notation. For example, the number 2.60×10^3 contains three significant figures and the number 2.600×10^3 contains four significant figures.

Rules for determining if a reported digit is significant are as follows.

1. Nonzero digits are always considered to be significant.
2. Zeros to the left of the first nonzero digit are never significant.
3. Zeros between nonzero digits are always significant.
4. Zeros at the right end of a number and to the right of the decimal are significant.
5. Zeros at the right end of a whole number are uncertain. Whole numbers should be reported in scientific or engineering notation to clarify the significant figures.

PROCEDURE:

1. Many of the dials and controls of laboratory instruments are labeled with metric prefixes. Check the controls on instruments at your lab station for metric prefixes. For example, check the SEC/DIV control on your oscilloscope. This control usually has more than one metric prefix associated with the switch positions. Meters are also frequently marked with metric prefixes. Look for others and list the instrument, control, metric

prefix, and its meaning in Table 1–3 of your report. The first line of Table 1–3 has been completed as an example.

2. The actual sizes of several electronic components are shown in Figure 1–1. Measure the quantities shown with a bold letter using a metric ruler. In Table 1–4, record the length in millimeters of each lettered quantity. Then change the measured length into the equivalent length in meters and record your results in Table 1–4. The first line of the table has been completed as an example.

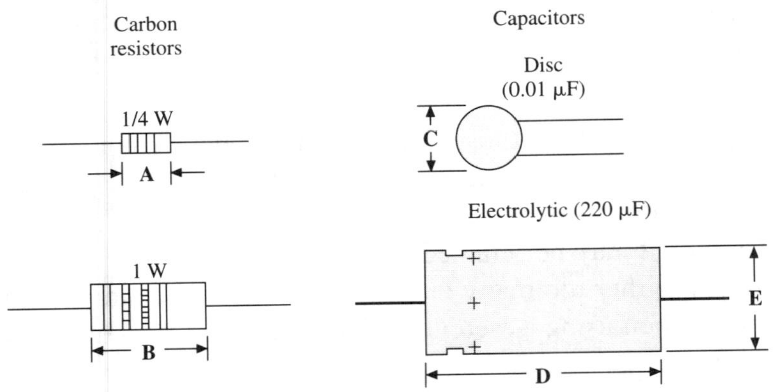

Figure 1–1

3. Rewrite the numbers in Table 1–5 of the report in scientific notation, engineering notation, and using one of the engineering metric prefixes. The first line has been completed as an example.

4. Convert the metric values listed in Table 1–6 into engineering notation. The first line has been completed as an example.

5. Using the rules given in the Summary of Theory for significant digits, determine the number of significant digits for each number listed in Table 1–7. Underline the significant digits and cite the pertinent rule number from the rules given in the Summary of Theory. The first three are completed as examples.

FOR FURTHER INVESTIGATION:

Metric prefixes are useful for solving problems without having a key in the exponent in your calculator. For example, when a milli prefix (10^{-3}) is multiplied by a kilo prefix (10^{+3}), the metric prefixes cancel, and the result has only the unit of measure. As you become proficient with these prefixes, math operations can be simplified and fewer keystrokes are required in solving the problem with a calculator. To practice this, determine the metric prefix for the answer when the operation indicated in Table 1–8 is performed. The first line is shown as an example.

APPLICATION PROBLEM:

A *schematic* is a shorthand drawing for showing the parts and electrical connections in a circuit. Examples of schematic symbols are shown in Figure 1–2.

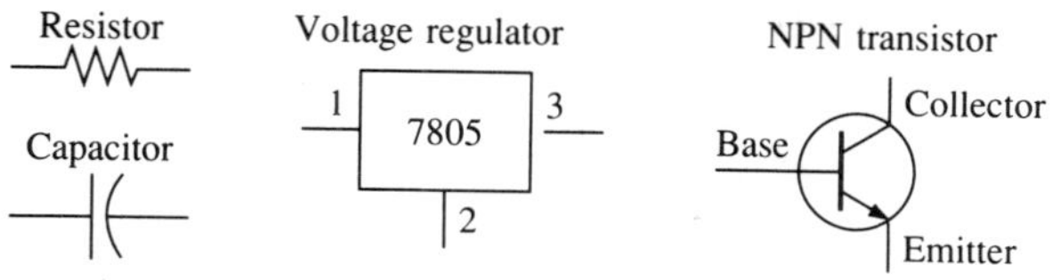

Figure 1–2

A comparison of the physical layout and the electrical schematic for a simple circuit is shown in Figure 1–3(a) and (b). The circuit shown is a regulator frequently used in power supplies.

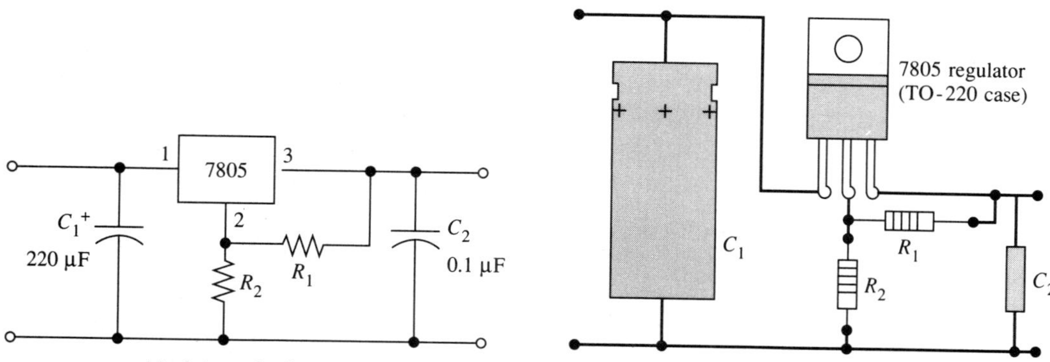

(a) Schematic diagram (b) Physical layout

Figure 1–3

Another basic circuit is the amplifier illustrated in the schematic in Figure 1–4. Using the example as a reference, draw the physical layout of the circuit shown in Figure 1–4. All resistors are ¼ W. Submit your drawing with your report.

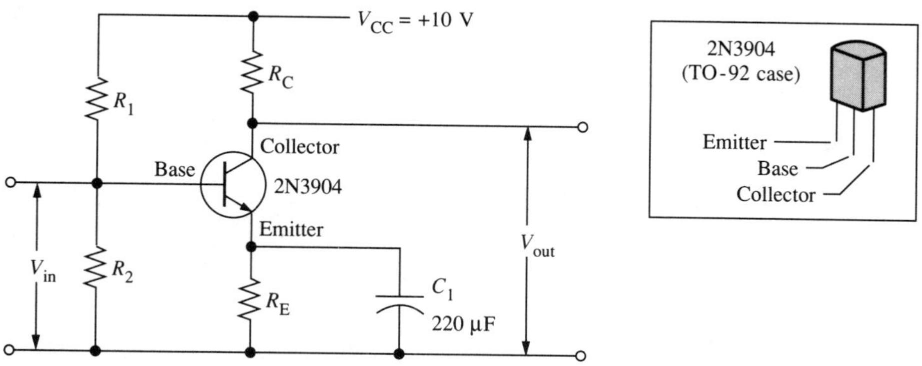

Figure 1–4

Report for Experiment 1

Name ____________
Date ____________
Class ____________

OBJECTIVES:

PROCEDURE:

DATA:

Table 1–3

Instrument	Control	Metric Unit	Meaning
Oscilloscope	SEC/DIV	ms	10^{-3} seconds

Table 1–4

Dimension	Length in millimeters	Length in meters
A	6.4 mm	6.4×10^{-3} m
B		
C		
D		
E		
F		
G		

Table 1–5

Number	Scientific Notation	Engineering Notation	Metric Value
0.0829 V	8.29×10^{-2} V	82.9×10^{-3} V	82.9 mV
48,000 Hz			
2,200,000 Ω			
0.000 015 A			
7,500 W			
0.000 000 033 F			
270 000 Ω			
0.000 010 H			

Table 1–6

Metric Value	Engineering Notation
100 pF	100×10^{-12} F
12 kV	
85.0 μA	
50 GHz	
33 kΩ	
250 mV	
7.8 ns	
2.0 MΩ	

Table 1–7

Number	Rule Number	Number	Rule Number
1.472	1	0.008	
4.09×10^6	3	60.90	
0.00150	4	84×10^3	
0.00842		0.105	
5.0		1.00×10^{-9}	

RESULTS AND CONCLUSION:

FURTHER INVESTIGATION RESULTS:

Table 1–8

Metric Unit in Operand	Mathematical Operation	Metric Unit in Operand		Metric Unit in Result
milli	multiplied by	milli	=	micro
kilo	multiplied by	micro	=	
nano	multiplied by	kilo	=	
milli	multiplied by	mega	=	
micro	divided by	nano	=	
micro	divided by	pico	=	
pico	divided by	pico	=	
milli	divided by	mega	=	

APPLICATION PROBLEM RESULTS:

EVALUATION AND REVIEW QUESTIONS:

1. For each metric prefix and unit shown, write the abbreviation of the metric prefix with the unit symbol.
   (a) kilowatt
   (b) milliampere
   (c) picofarad
   (d) nanosecond
   (e) megohm
   (f) microhenry
2. Write the metric prefix and unit name for each of the abbreviations shown.
   (a) MW
   (b) nA
   (c) μJ
   (d) mS
   (e) kΩ
   (f) GHz
3. Using your calculator, perform the following multiplications. Express your answer in scientific notation.
   (a) $(3.6 \times 10^4)(8.8 \times 10^{-4})$
   (b) $(-4.0 \times 10^{-6})(2.7 \times 10^{-1})$
   (c) $(-7.5 \times 10^2)(-2.5 \times 10^{-5})$
   (d) $(56 \times 10^3)(9.0 \times 10^{-7})$
4. Using your calculator, perform the following divisions. Express your answer in engineering notation.
   (a) $\dfrac{(4.4 \times 10^9)}{(-7.0 \times 10^3)}$
   (b) $\dfrac{(3.1 \times 10^2)}{(41 \times 10^{-6})}$
   (c) $\dfrac{(-2.0 \times 10^4)}{(-6.5 \times 10^{-6})}$
   (d) $\dfrac{(0.0033 \times 10^{-3})}{(-15 \times 10^{-2})}$
5. For each result in problem 4, change the answer to one with a metric prefix.
   (a)

   (b)

   (c)

   (d)

6. Explain why the reported value 10.0 V conveys more information about a measured quantity than the value 10 V.

2 Laboratory Meters and Power Supply

OBJECTIVES:

After performing this experiment, you will be able to:

1. Read analog meter scales including multiple and complex scales.
2. Operate the power supply at your lab station.
3. Explain the functions of the controls for the multimeter at your lab station. Use it to make a voltage reading.

READING:

Floyd, *Principles of Electric Circuits,* Sections 2–1 through 2–7
Operator's Manual for Laboratory Multimeter and Power Supply

MATERIALS NEEDED:

None
For Further Investigation: Meter calibrator
Application Problem: Eight different colored wires

SUMMARY OF THEORY:

The measurement of various electrical quantities is basic to determining circuit performance. These quantities include voltage, current, power, frequency, and many others. Many electrical quantities are measured with meters. The schematic symbol for a meter is shown in Figure 2–1. The meter function is shown on the schematic with a letter or symbol. One type of meter is the *multimeter,* an instrument that combines three basic instruments to measure either resistance, voltage, or current. Multimeters may be either *analog* or *digital* and may contain either active or passive circuitry. Analog multimeters use a pointer to indicate on a numbered scale the value of the measured quantity. One example of an analog meter is the VOM (for volt-ohm-milliammeter). A digital multimeter (or DMM) shows the measured quantity as a number. The digital multimeter is rapidly replacing analog multimeters because of superior performance and ease of use. Examples of a portable VOM and DMM are shown in Figure 2–2.

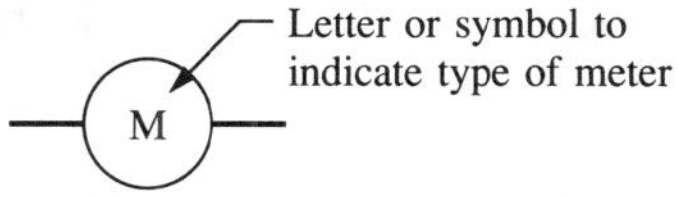

Figure 2–1
Meter Symbol

For this experiment manual, it is assumed that you have a DMM available at your lab station and all references to the lab station meter are to your DMM. All experiments, including this one, can be performed with an analog meter if a DMM is not available.

Electrical quantities extend from the very small to the very large. Resistance, for example, can vary from less than 1 Ω to more than 1,000,000 Ω. Meters must have some means of accommodating these numbers. The position of the decimal place is determined by the *range* switch on the meter. The user selects an appropriate range to display the measured

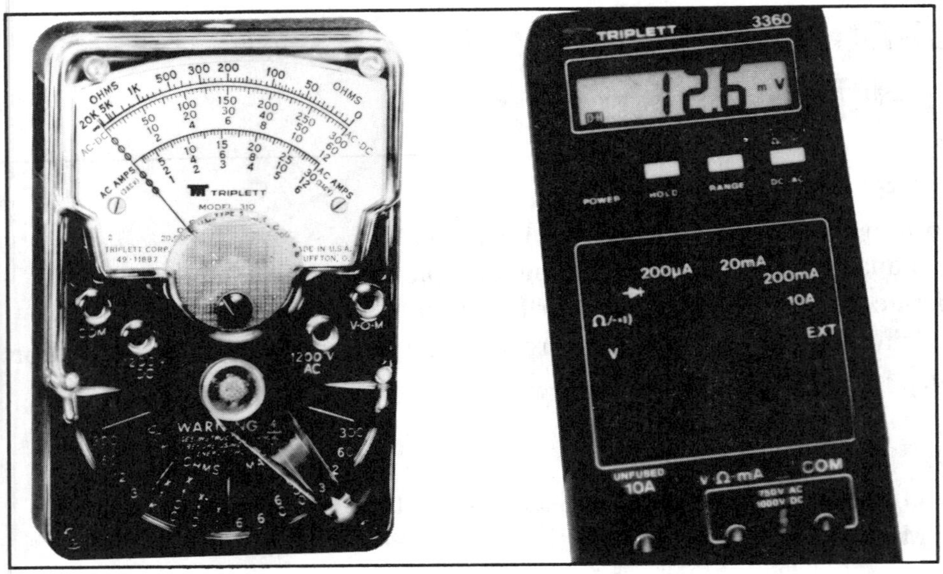

Figure 2–2

number. Some meters have *autoranging,* which means they can change ranges automatically. An autoranging meter may also have an AUTO/HOLD switch, which allows the meter to either operate in the autoranging mode or to hold the last range setting. To operate an autoranging meter, the meter will normally be used in the AUTO mode. The function to be measured is selected and the meter is connected to the circuit under test. The user must be careful to connect the meter correctly for the measurement to be made. Examples of how to connect an autoranging DMM for measurement of voltage and resistance are shown in Figure 2–3.

Current measurements require special care to avoid damage to the meter. <u>A current meter must never be connected across a voltage source.</u> Current measurements are

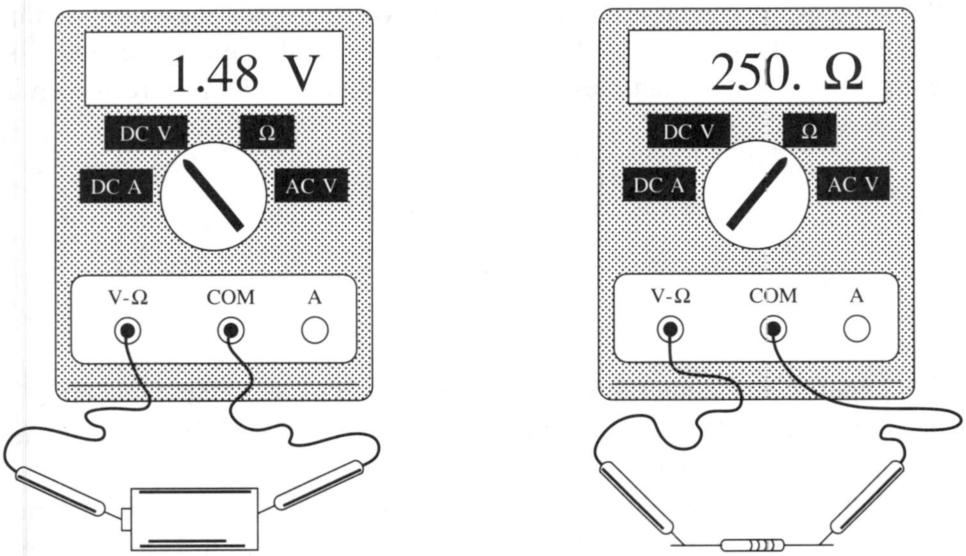

(a) DC voltage measurement

(b) Resistance measurement

Figure 2–3

accomplished by selecting current (either ac or dc) with the function select switch and moving the probes to the AMPS socket. In this experiment, current will not be measured.

Many electronic measurements are made with analog meters. Analog meters can be calibrated to read almost any physical quantity, including voltage, current, power, or even nonelectrical quantities such as weight, speed, or light. The scales on analog meters may be either linear or nonlinear. There may be several scales on the same meter face.

A linear meter scale is marked in equally spaced divisions across the face of the meter. Figure 2–4 shows a linear meter scale. The major divisions, called *primary* divisions, are usually numbered. Between the primary divisions are smaller divisions called *secondary* divisions. To read this scale, note the number of secondary divisions between the numbered primary divisions and determine the value of each secondary division. The scale shown has 9 secondary marks between the primary marks. If we include one primary mark, there are 10 subdivisions. Each primary division is an increment of 5.0 milliamperes (mA), making each secondary division worth 0.5 mA. Frequently a meter is used for several ranges. The meter shown in Figure 2–4 could, for example, have a 200 mA full-scale range. It is up to the user to then set the decimal place, depending on which range has been selected. If the user selects the 200 mA full-scale range, then there are 50 mA between each primary division.

Usually, meters with more than one range have several scales called *multiple* scales. A meter with two scales is illustrated in Figure 2–5. Each scale can represent one or more ranges. In this case, the user must choose the appropriate scale *and* set the decimal place.

Figure 2–4

Figure 2–5

The top scale in Figure 2–5 has a full-scale value of 10 V. This scale should be read if the 10 V range is selected. For this scale, the primary divisions are 2 V and the secondary divisions are 0.5 V. If this range is selected, the meter reading is then interpreted as 8.5 V.

VOMs and some instruments contain meters that can be used for more than one function. These scales are called *complex* scales. To read a complex scale, the user chooses the appropriate scale based on the function *and* the range selected. Figure 2–6 shows a complex scale from a VOM. Notice that the top scale is used for measuring resistance and is nonlinear. Although the meter shown indicates increasing resistance toward the right-hand side, on many meters this is reversed.

If the function selected is resistance, then the top scale is selected. Notice that the secondary divisions change values across the scale. To determine the reading, the primary divisions on each side of the pointer are noted. The secondary divisions can then be assigned values by counting the number of secondary divisions between the primary marks.

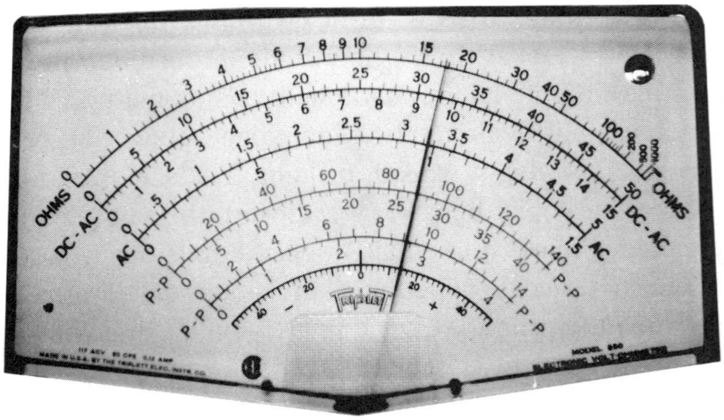

Figure 2–6

PROCEDURE:

1. Observe the meters shown in Figure 2–7(a) and (b) of the report. These are linear meters with one range. Determine the value of the primary and secondary divisions for each meter. Then determine the meter reading. Record your observations in the space provided in the report.

2. The meter shown in Figure 2–8 of the report is a VOM scale. Answer the questions in the report for this meter.

3. Look at the meter on the power supply at your lab station. Some power supplies have meters that monitor either voltage or current. There may be more than one range or several supplies built into the same chassis, so the meter may have multiple or complex scales. Review the meter and meter controls and answer the questions in the report about your power supply meter.

4. Review the controls for the power supply at your lab station. The operator's manual is a good resource if you are not sure of the purpose of a control. In the space provided in the report, describe the features of your supply (multiple outputs, current limiting, tracking, etc.).

5. In this step, you will set the power supply for a specific voltage and measure that voltage with your laboratory meter. Review the operator's manual for the DMM (or VOM) at your lab station. Review each control on the meter. Then select +DC and VOLTS on the DMM. If your DMM is not autoranging, select a range that will measure +5.0 V dc. The best choice is a range that is the smallest range larger than +5.0 V. Connect the test leads together and verify that the reading is zero. (*Note:* A digital meter may have a small digit in the least significant place.)

6. Turn on the power supply at your station and use the meter on the supply to set the output to +5.0 V. Then use the DMM to confirm that the setting is correct. Record the readings of the power supply meter and the DMM in Table 2–1.

7. Set the output to +12.0 V and record the readings of the power supply meter and the DMM in Table 2–1.

8. Set the supply to the minimum output voltage and record the readings of the power supply meter and the DMM in Table 2–1.

FOR FURTHER INVESTIGATION:

The sensitivity of a panel meter is a number that describes how much current is required to obtain full-scale deflection from the meter. Meter sensitivity is easily determined with a *meter calibrator.* If you have a meter calibrator available, go over the operator's manual and learn how to find the full-scale current through an inexpensive panel meter. Then obtain a small panel meter and measure its sensitivity.

Calibration is the process of comparing a device with a higher-accuracy standard and then adjusting it or recording the difference. Set the meter calibrator for 25%, 50%, and 75% of the full-scale current. Record the difference between the reading on the calibrator and the reading on the meter. Graph the meter current versus the calibrator current on Plot 2–1 of your report. This graph is called a calibration curve for the meter.

APPLICATION PROBLEM:

A continuity tester is an instrument used to determine if there is a low-resistance path between two points. An ohmmeter can be used as a continuity tester to determine if there is a path for current between conductors. If the ohmmeter reads nearly 0 Ω, the conductors are connected; if no reading is obtained, the conductors are open. Some meters have a built in audio signal to indicate when there is continuity.

Obtain eight different-colored wires. Have a lab partner connect the ends of random pairs of wires so that you have four pairs of wires. Tape the ends so they are concealed. Using your ohmmeter as a continuity tester, design a test procedure that determines which pairs are connected together in the fewest number of steps. Summarize your procedure in the report.

Report for Experiment 2

Name ______________________
Date ______________________
Class ______________________

OBJECTIVES:

DATA:

Step 1:

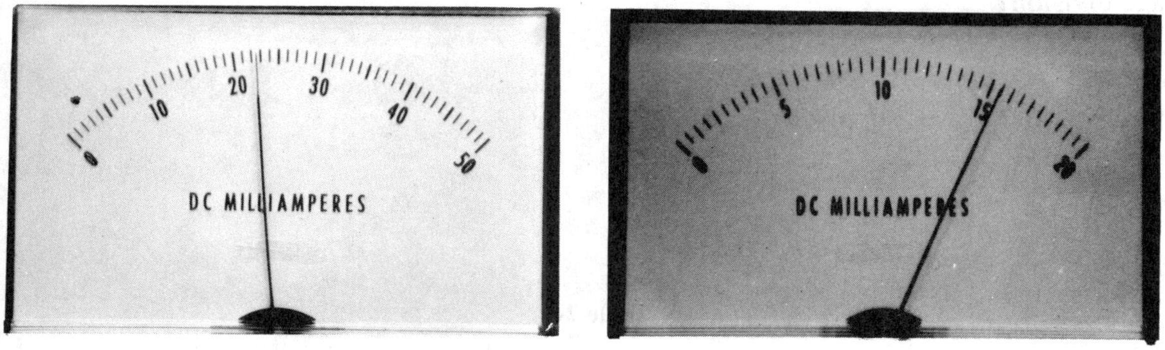

(a) (b)

Figure 2–7

Meter Readings:

Figure 2–7(a) ______________________ Figure 2–7(b) ______________________

Step 2:

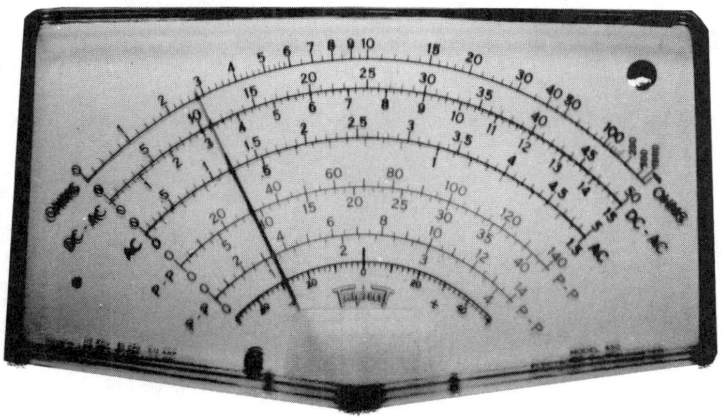

Figure 2–8

For the meter in Figure 2–8, assume the OHMS function is selected and the range selected is X 1K ohms. What does the meter indicate for a resistance? ______________________

What is the meter reading if the range selected is the 50 V DC range? ____________

What is the meter reading if the range selected is the 15 V AC range? ____________

Step 3:
Is the meter on your power supply used for more than one function? ________ If so, what determines which function is monitored? ____________________________

Does it have multiple scales? ________ complex scales? ________

What is the smallest primary voltage division? ________ the smallest secondary voltage division? ________

Step 4:
Features of your power supply:

Table 2–1

Step Number	Power Supply Meter Reading	Lab Station Meter Reading
6		
7		
8		

RESULTS AND CONCLUSION:

FURTHER INVESTIGATION RESULTS:

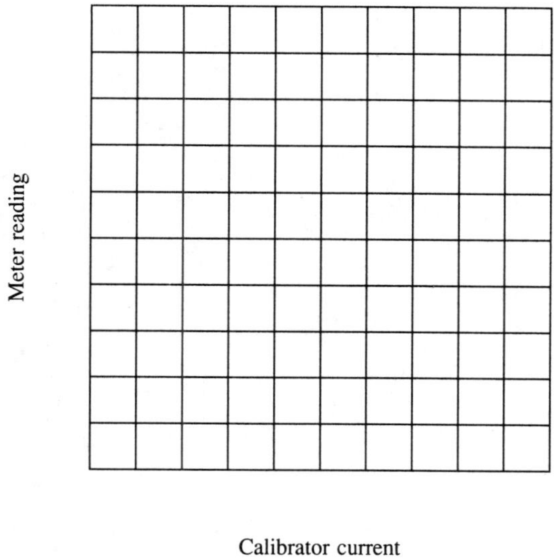

Plot 2–1

APPLICATION PROBLEM RESULTS:

Test Procedure:

EVALUATION AND REVIEW QUESTIONS:

1. Compare the precision of the power supply voltmeter with the DMM or VOM at your lab station. Does one meter have an advantage for measuring 5.0 V? Explain your answer.

2. What is meant by an autoranging meter? What type is at your lab station?

3. What is the difference between a multiple scale and a complex scale?

4. What is the difference between a linear scale and a nonlinear scale?

5. Assume a scale has four secondary marks between the primary marks numbered 3.0 and 4.0. If the pointer is on the first mark, what is the reading on the meter?

6. What special precaution is necessary for measuring current with a meter?

3 Measurement of Resistance

OBJECTIVES:

After performing this experiment, you will be able to:

1. Determine the listed value of a resistor using the resistor color code.
2. Use the DMM (or VOM) to measure the value of a resistor.
3. Determine the percent difference between the measured and listed values of a resistor.
4. Measure the resistance of a potentiometer and explain its operation.

READING:

Floyd, *Principles of Electric Circuits,* Sections 2–5 through 2–8

MATERIALS NEEDED:

Resistors: Ten assorted values

One potentiometer (any value)

SUMMARY OF THEORY:

Resistance is the opposition a substance offers to current flow. The unit for resistance is the *ohm,* symbolized with the Greek letter capital omega (Ω). A resistor is a component designed to have a specific resistance and wattage rating. Resistors limit current but in doing so, produce heat. The physical size of a resistor is related to its ability to dissipate heat, *not* to its resistance. A physically large resistor can dissipate more heat than a smaller resistor; hence the larger one would have a higher wattage rating than the smaller one.

Resistors are either fixed (constant resistance) or variable. Fixed carbon and film resistors are usually color-coded with a four-band code that indicates the specific resistance and tolerance. Each color stands for a number as described in Floyd's text and reprinted in Table 3–1 for convenience. Figure 3–1 shows how to read the resistance and tolerance of a four-band resistor.

Table 3–1

	Digit	Color
Resistance value, first three bands	0	Black
	1	Brown
	2	Red
	3	Orange
	4	Yellow
	5	Green
	6	Blue
	7	Violet
	8	Grey
	9	White
Tolerance, fourth band	5%	Gold
	10%	Silver
	20%	No band

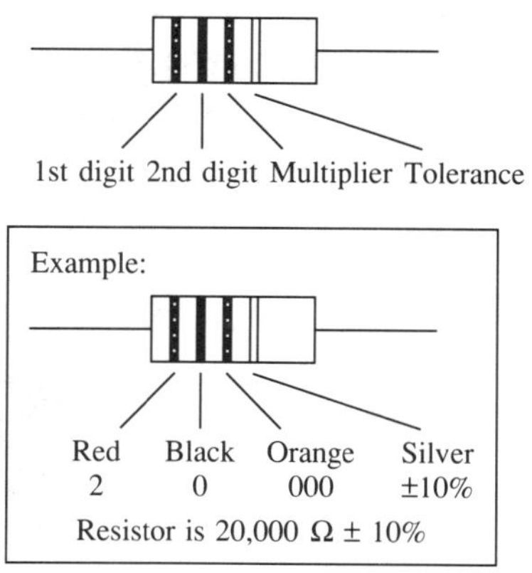

Note: In the multiplier band, Gold = X 0.1
Silver = X 0.01

Figure 3–1

The resistance of resistors is measured using a DMM or VOM as described in Experiment 2. If you are using a VOM, the zero reading should be checked whenever you change ranges on the meter by touching the test leads together. If you are using a nonautoranging DMM, a suitable range needs to be selected. Resistance normally should not be measured in a circuit, because other resistors in the circuit will affect the reading. The resistor to be measured is removed from the circuit and the test leads are connected across the resistance. The resistor under test should not be held between the fingers because body resistance can affect the reading, particularly with high-value resistors. (It is okay to hold one end of the resistor under test.)

The most common form of variable resistor is the potentiometer. The potentiometer is a three-terminal device with the outer terminals having a fixed resistance between them and the center terminal connected to a moving contact. The moving contact is connected to a shaft that is used to vary the resistance between the moving contact and the outer terminals. Potentiometers are commonly found in applications such as volume controls.

PROCEDURE:

1. Obtain 10 four-band fixed resistors. Record the colors of each resistor in Table 3–2. Use the resistor color code to determine the color-code resistance of each resistor. Then measure the resistance of each resistor and record the measured value in Table 3–2 in the report. The first line has been completed as an example.

2. Compute the percent difference between the measured and color-coded values using the equation:

$$\%\ \text{diff} = \frac{R_{\text{measured}} - R_{\text{color code}}}{R_{\text{color code}}} \times 100$$

Enter the computed differences in Table 3–2.

3. Obtain a potentiometer. Number the terminals 1, 2, and 3, as illustrated in Figure 3–2. Vary the potentiometer's shaft while you monitor the resistance between terminals 1 and 2. Notice that the resistance between the outside terminals does not change as the shaft is varied. Record the resistance between terminals 1 and 2 of the potentiometer (the outside terminals) in Table 3–3 of your report.

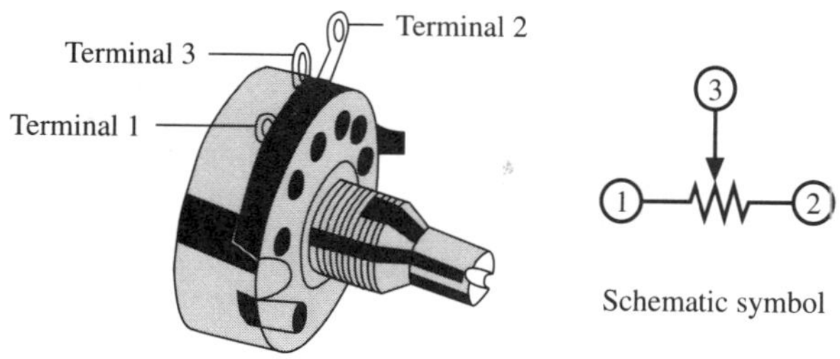

Figure 3–2

4. Turn the potentiometer completely counterclockwise (CCW). Measure the resistance between terminals 1 and 3. Then measure the resistance between terminals 3 and 2. Record the measured resistances in Table 3–3. Compute the sum of the two readings and enter it into Table 3–3.

5. Turn the shaft 1/3 turn clockwise (CW) and repeat the measurements in step 4.

6. Turn the shaft 2/3 turn CW and repeat the measurements in step 4.

FOR FURTHER INVESTIGATION:

This experiment described how to read 5% and 10% tolerance resistors. The same idea is used for most 1% and 2% resistors except that 1% and 2% will have one more color band than 5% and 10% resistors. The first three bands represent the first, second, and third significant figures. The fourth band represents the multiplier band. The decimal point is assumed to be after the third significant figure and then moved by the amount shown in the multiplier band. The fifth band represents the tolerance band. A 1% resistor has a brown tolerance band and a 2% resistor has a red tolerance band. There is a space between the fourth and fifth band to avoid mistaking the tolerance band for the first significant figure and mistakenly reading the resistor backwards. For each of the resistors shown in Table 3–4, find the remaining information and complete the table. The first line is completed as an example.

APPLICATION PROBLEM:

At room temperature, all known materials have some resistance. Usually the resistance of wire can be ignored, but not always. Sometimes, when it is necessary to obtain a very small resistance, the resistance of wire is used. Each substance has a specific resistivity, which is characteristic of the material. The resistivity of copper, for example is $1.69 \times 10^{-8}\ \Omega \cdot \text{m}$. The resistance of a wire depends on its resistivity, length, and its cross sectional area as given in the equation:

$$R = \frac{\rho l}{A}$$

where R = resistance in ohms
ρ = resistivity, in ohm-meters
l = length of the wire in meters
A = area of wire in square meters

Assume you need a 0.5 Ω resistor and you have available #22 gauge wire with a diameter of 6.38×10^{-4} m. Compute the length of wire needed for the resistor. If you have a means of measuring the result, cut a piece based on your calculation and report the measured result.

Report for Experiment 3

Name ______________
Date ______________
Class ______________

OBJECTIVES:

PROCEDURE:

DATA:

Table 3–2

Resistor	Color of Band				Color-Code Value	Measured Value	% Difference
	1st	2nd	3rd	4th			
0	brown	green	red	silver	1.5 kΩ ± 10%	1.46 kΩ	−2.7%
1							
2							
3							
4							
5							
6							
7							
8							
9							
10							

Table 3–3

Step 3: Total resistance between terminals 1 and 2 = ______________

Step	Shaft Position	Resistance Measured Between:		Sum of Resistance Readings
		Terminals 1–3	Terminals 3–2	
4				
5				
6				

RESULTS AND CONCLUSION:

FURTHER INVESTIGATION RESULTS:

Table 3–4

Resistor	Color of Band					Color-Code Value	Minimum Value	Maximum Value
	1st	2nd	3rd	4th	5th			
0	brown	red	violet	brown	red	1.27 kΩ ± 2%	1.24 kΩ	1.30 kΩ
1						536 Ω ± 1%		
2						3.30 MΩ ± 2%		
3						97.6 kΩ ± 1%		
4	violet	green	black	gold	red			
5	brown	green	black	black	brown			

APPLICATION PROBLEM RESULTS:

EVALUATION AND REVIEW QUESTIONS:

1. (a) Identify any of the resistors measured in Table 3–2 that are out of tolerance.

   (b) You suspect that the percent difference between color-coded and measured values could be due to error in the meter. How could you find out if you are correct?

2. Predict the resistance between terminals 1–3 and 3–2 for the potentiometer if the shaft is rotated fully CW.

3. Determine the resistor color code for the following resistors. The tolerance is 10%.
   (a) 12 Ω ____________________
   (b) 6.8 kΩ ____________________
   (c) 910 Ω ____________________
   (d) 4.7 MΩ ____________________
   (e) 1.0 Ω ____________________

4. Determine the expected value for resistors with the following color codes.
   (a) red-red-black-gold ____________
   (b) violet-green-brown-silver ____________
   (c) green-brown-brown-gold ____________
   (d) white-brown-gold-gold ____________
   (e) grey-red-yellow-silver ____________

5. A resistor is color-coded red-violet-orange-gold.
   (a) What is the largest value the resistor can be and still be in tolerance?

   (b) The smallest value?

6. Explain why experimental calculations should use measured values of resistors rather than color-coded values:

TECH TIP REVIEW:
Each of the following circuit problems relate to the TECH TIP circuit shown in Figure 2–46 of the Floyd text. Can you determine the most probable cause for each problem?

1. The lamp does not light; a voltage measurement reveals no voltage across the lamp but +12 V across the fuse.

2. The circuit blows fuses even when the bulb is not in the circuit.

3. The light is dim and the variable resistor does not change the brightness. The potentiometer is replaced but the trouble persists.

4 Ohm's Law

OBJECTIVES:

After performing this experiment, you will be able to:

1. Measure and plot the current-voltage relationship for a resistor.
2. Construct a graph of the data from objective 1.
3. Given a graph of current-voltage for a resistor, determine the resistance.

READING:

Floyd, *Principles of Electric Circuits,* Sections 3–1 through 3–6

MATERIALS NEEDED:

Resistors:

One 1.0 kΩ, one 1.5 kΩ, one 2.2 kΩ

One dc ammeter, 0–10 mA

For Further Investigation: One 5 V zener diode

SUMMARY OF THEORY:

The flow of electrical charge in a circuit is called *current*. Current is measured in units of *amperes,* or amps for short. The ampere is defined as one coulomb of charge moving past a point in one second. Current is abbreviated with the letter *I* (for *Intensity*) and is frequently shown with an arrow to indicate the direction of flow. Conventional current is defined as the direction a positive charge would move under the influence of an electric field. When electrons move, the direction is opposite to the direction defined for conventional current. To clarify the difference, the term *electron flow* is frequently applied to current in the opposite direction of conventional current flow. The experiments in this lab book work equally well with either definition.

The relationship between current and voltage specify the characteristics of an electrical device. One convenient way to represent these quantities is with a graph. In order to construct the graph, one of the variables is changed and the response of the other variable is observed. The variable that was initially moved is called the independent variable; the one that responds is called the dependent variable. By convention, the independent variable is plotted along the *x*-axis (the horizontal axis) and the dependent variable is plotted along the *y*-axis (the vertical axis). In this experiment, the voltage will be controlled (independent) and the current will respond (dependent).

Fixed resistors have a straight-line, or *linear,* current-voltage curve. This linear relationship illustrates the basic relationship of Ohm's law—namely, that the current is proportional to the voltage for constant resistance. Ohm's law is the most important law of electronics. It is written in equation form as

$$V = IR$$

where V represents voltage, I represents current, and R represents resistance.

PROCEDURE:

1. Measure three resistors with listed values of 1.0 kΩ, 1.5 kΩ, and 2.2 kΩ. Record the measured values in Table 4–1.

2. Connect R_1 into the circuit shown in Figure 4–1. The schematic diagram and an example of the protoboard wiring is shown.

Caution! Current meters can be easily damaged if they are incorrectly connected. Have your instructor check your connections before applying power.

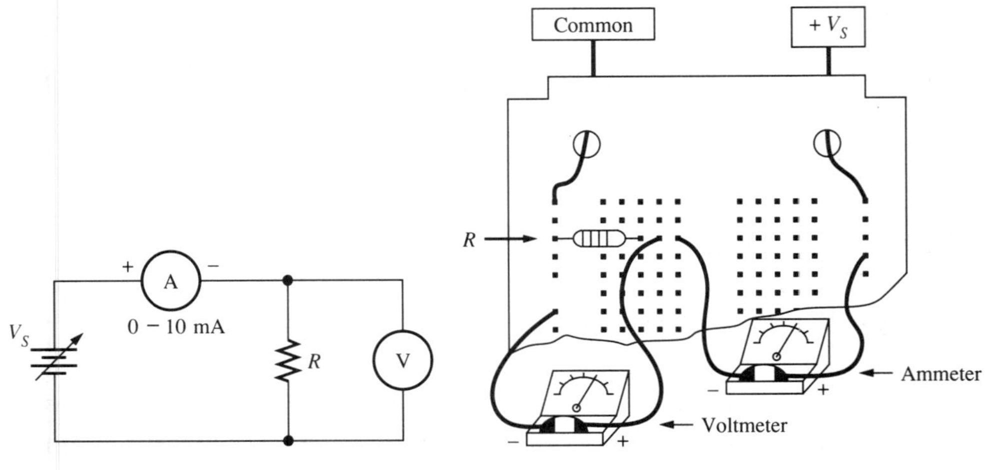

Figure 4–1

3. Adjust the power supply for a voltage of 2.0 V. Measure the current that is flowing through the resistor and record it in Table 4–2.

4. Adjust the power supply for 4.0 V and measure the current. Record the current in Table 4–2. Continue taking current readings for each of the voltages listed in Table 4–2.

5. Replace R_1 with R_2 and repeat steps 3 and 4. Record the data in Table 4–3.

6. Replace R_2 with R_3 and repeat steps 3 and 4. Record the data in Table 4–4.

7. On Plot 4–1, graph all three *I-V* curves using the data from Tables 4–2, 4–3, and 4–4. Plot the dependent variable (current) on the *y*-axis and the independent variable (voltage) on the *x*-axis. Choose a scale for the graph that spreads the data over the entire grid. Label the three resistance curves with the resistor value.

FOR FURTHER INVESTIGATION:

Not all devices have a linear current-voltage relationship. (This is what makes electronics interesting!) Investigate a zener diode *I-V* curve. The circuit is shown in Figure 4–2. The 1.0 kΩ resistor is used to limit the total current in the circuit. Notice the polarity of the zener diode. Measure the voltage across the zener diode as the power supply is varied. The circuit is a series circuit so the zener current is the same as the current read by the ammeter. Summarize your results with a graph of the *zener current* as a function of the *zener voltage*.

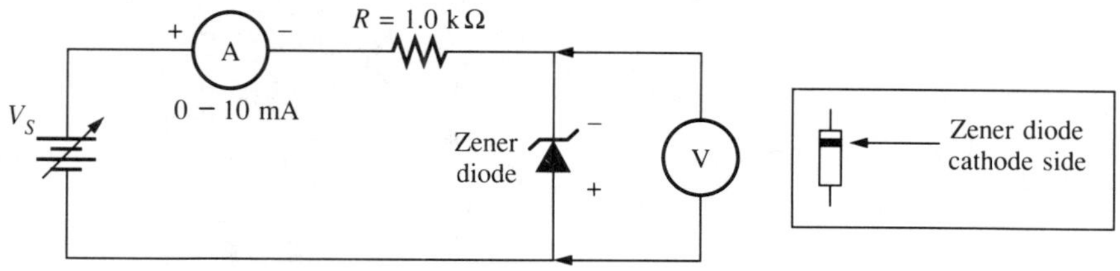

Figure 4–2

APPLICATION PROBLEM:

A student decides to set up switches for testing the resistors in this experiment. The student has only two double-pole, single-throw switches. A partial schematic is drawn in Figure 4–3 of the report. The two switches are to be wired in a way that connects only one resistor at a time to the voltage source. When S_1 is in position **B,** only R_1 is in the circuit; when S_1 is moved to position **A** and S_2 is in position **C,** only R_3 will be in the circuit; when S_2 is now moved to position **D,** only R_2 is in the circuit. Complete the schematic shown in Figure 4–3.

Report for Experiment 4

Name ____________________
Date ____________________
Class ____________________

OBJECTIVES:

PROCEDURE:

DATA:

Table 4–1

Component	Listed Value	Measured Value
R_1	1.0 kΩ	
R_2	1.5 kΩ	
R_3	2.2 kΩ	

Table 4–2 (R_1)

$V_s =$	2.0 V	4.0 V	6.0 V	8.0 V	10.0 V
$I =$					

Table 4–3 (R_2)

$V_s =$	2.0 V	4.0 V	6.0 V	8.0 V	10.0 V
$I =$					

Table 4–4 (R_3)

$V_s =$	2.0 V	4.0 V	6.0 V	8.0 V	10.0 V
$I =$					

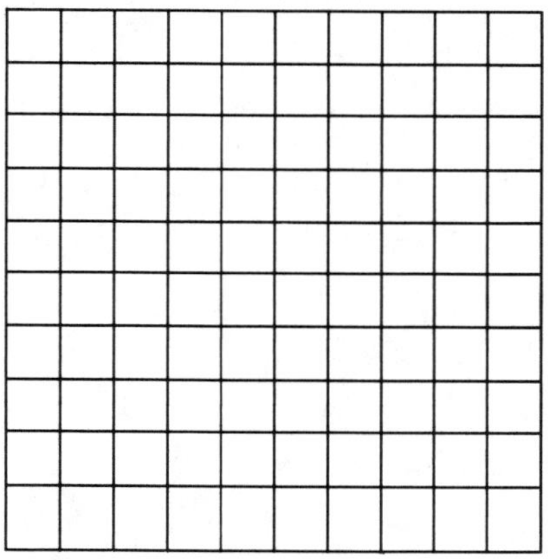

Plot 4–1

RESULTS AND CONCLUSION:

FURTHER INVESTIGATION RESULTS:

Table 4–5 (Zener diode)

$V_s =$	2.0 V	4.0 V	6.0 V	8.0 V	10.0 V
$V_z =$					
$I_z =$					

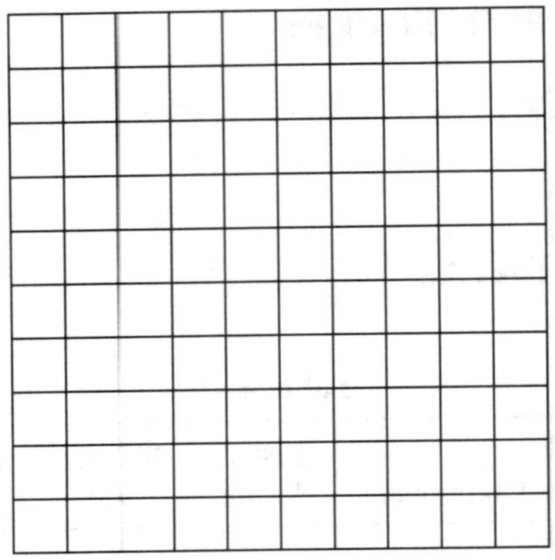

Plot 4–2

APPLICATION PROBLEM RESULTS:

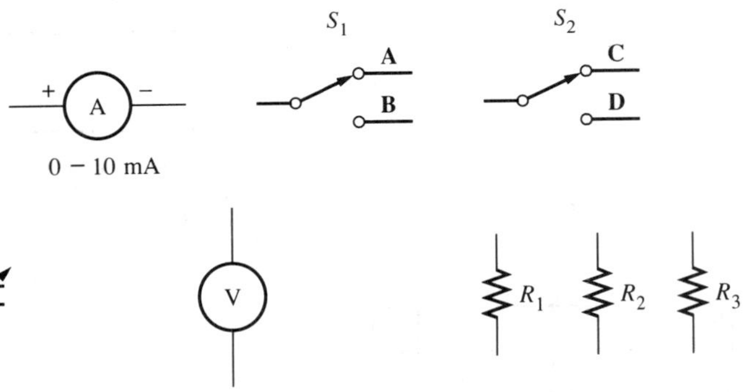

Figure 4–3

EVALUATION AND REVIEW QUESTIONS:

1. The slope of a line is the change in the y direction divided by the change in the x direction. The definition for slope is illustrated in Figure 4–4. Find the slope for each resistor on Plot 4–1. Note that the slope has units for conductance, the siemens.

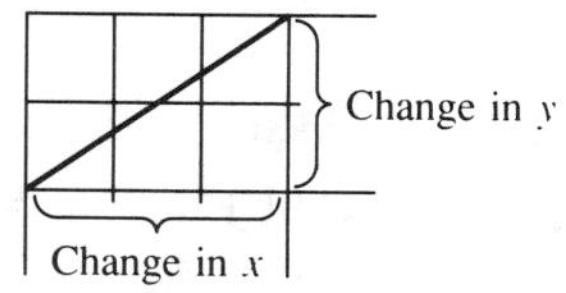

$$\text{Slope} = \frac{\text{Change in } y}{\text{Change in } x} = \frac{2}{3}$$

Figure 4–4

2. What happens to the slope of the I-V curve for larger resistors?

3. (a) If the resistance is halved and the voltage is not changed, what will happen to the current in a resistive circuit?

   (b) If the voltage is doubled and the resistance is not changed, what will happen to the current in a resistive circuit?

4. If the current in a resistive circuit is 24 mA and the applied voltage is 48 V, what is the resistance?

5. What current will flow through a 10 Ω resistor with 5.0 V applied?

6. The resistance of a tungsten bulb increases as it gets hotter. How does this help explain why many bulbs burn out when they are first turned on?

TECH TIP REVIEW:
Each of the following problems relate to the TECH TIP resistor box shown in Figure 3–15 of the Floyd text. Assume the resistor box is connected in series with an ammeter and a power supply. The power supply is set to 2.0 V. Determine the probable cause of each of the following problems.

1. Assume the resistor box is set to the R2 position. A voltmeter measurement reveals that there are 2.0 V across the terminals of the resistor box and the ammeter reads 2.0 mA.

2. Assume the resistor box is set to the R3 position. A voltmeter measurement reveals that there are 2.0 V across the terminals of the resistor box and the ammeter reads zero.

3. Assume the resistor box is set to the R4 position. The ammeter reads zero. A voltmeter measurement reveals that 2.0 V is across the ammeter and no voltage is across the resistor box.

5 Power in DC Circuits

OBJECTIVES:
After performing this experiment, you will be able to:
1. Find the power in a resistor using each of the three basic power equations.
2. Indirectly measure the power in a variable resistor at various settings of resistance.
3. Plot the power dissipated as a function of resistance for the variable resistor of objective 2.

READING:
Floyd, *Principles of Electric Circuits,* Sections 4–1 through 4–6

MATERIALS NEEDED:
One 2.7 kΩ resistor
One 10 kΩ potentiometer
One 0–10 mA ammeter
For Further Investigation: One LED

SUMMARY OF THEORY:
In physics, the unit for measuring energy or work is the joule. One joule is equivalent to the energy expended (or work accomplished) when a one newton weight (about 3.6 oz) is lifted over a distance of one meter. For electrical circuits, voltage represents the energy expended when a unit of positive charge is moved from one point to another point of higher potential. Alternatively, it is the energy released when one coulomb of charge moves from a point of higher potential to one of lower potential.

When current flows through a resistor, energy is dissipated as the current flows from the higher potential to the lower potential. In a resistance, the energy is dissipated in the form of heat. The power dissipated is defined as the energy given up per unit of time. Power in an electrical circuit is defined by the equation

$$P = IV$$

This equation says that energy dissipated per time is equal to the charge per time *(I)* times the energy per charge *(V)*. Power is measured in joules per second, which defines a unit called the **watt.** In an electrical circuit, one watt is the power developed when one volt is across one ohm of resistance. By applying Ohm's law to the defining law for power, two more useful equations for power can be found:

$$P = I^2R$$

and

$$P = \frac{V^2}{R}$$

The physical size of a resistor is related to the amount of heat it can dissipate. Therefore, larger resistors are rated for more power than smaller ones. Carbon composition resistors are available with standard power ratings ranging from 1/8 W to 2 W. For most typical low voltage applications (15 V or less and at least 1 kΩ of resistance) a 1/4 W resistor is satisfactory.

In this experiment, you will find the power dissipated in a fixed resistor and verify that each of the three preceding equations gives you the same result. Then you will find what happens to the power dissipated in a variable resistor as resistance is varied. Later, in Chapter 8, you will find out more about the results of this experiment.

PROCEDURE:

1. Measure the actual resistance of R_1. The color-coded value is 2.7 kΩ. Enter the measured resistance (in kΩ) in Table 5–1 and in the first column of Table 5–2.

2. Construct the circuit shown in Figure 5–1. The ammeter is connected in series, as in the previous experiment. Have your instructor check your circuit before applying power if you are not sure of your connections.

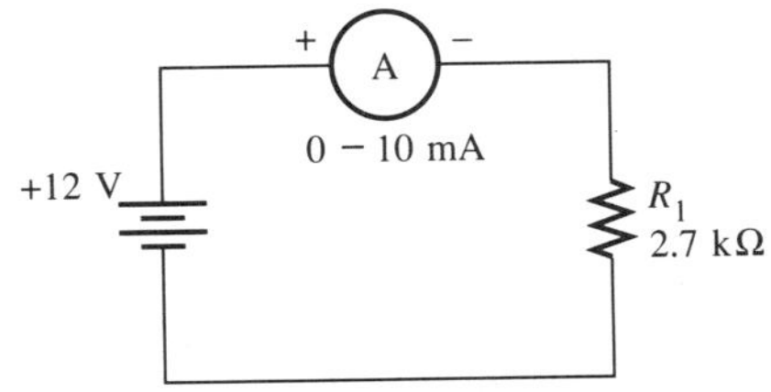

Figure 5–1

3. Using your DMM, set the power supply voltage to 12.0 V. This voltage also appears across R_1 (ignoring the very small voltage drop across the ammeter). Record the voltage and current (in mA) in the top portion of Table 5–2.

4. Using the measured resistance, voltage, and current, compute the power dissipated in R_1. Use each of the three forms of the power law and enter your results in the bottom portion of Table 5–2. You should find reasonable agreement between the three methods.

Determining Power in a Variable Resistance

5. Modify the circuit by removing the ammeter and adding a 10 kΩ potentiometer in series with R_1, as shown in Figure 5–2. R_2 is the 10 kΩ potentiometer. Connect the center (variable) terminal to one of the outside terminals. Use this and the remaining terminal as a variable resistor. Adjust the potentiometer for 0.5 kΩ. (Always remove the power source when measuring resistance).

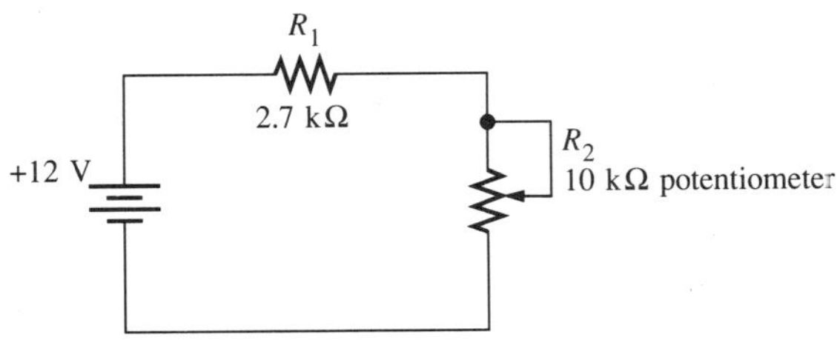

Figure 5–2

6. Measure the voltage across R_1 and the voltage across R_2. Enter the measured voltages in Table 5–3. As a check, make sure that the sum of V_1 plus V_2 is equal to 12.0 V. Then compute the power dissipated in R_2 using the equation

$$P_2 = \frac{(V_2)^2}{R_2}$$

Note that if the resistance is entered in kilohms, the power will be in milliwatts. Enter the computed power in Table 5–3.

7. Disconnect the power supply and set R_2 to the next value shown in Table 5–2. Reconnect the power supply and repeat the measurements made in step 2. Continue in this manner for each of the resistance settings shown in Table 5–3.

8. Using the data in Table 5–3, graph the relationship of the power, P_2, as a function of resistance R_2 on Plot 5–1. Since resistance is the independent variable, plot it along the *x*-axis and plot power along the *y*-axis. An *implied* data point can be plotted at the origin because there can be no power dissipated in R_2 without resistance. A smooth curve can then be drawn to the origin.

FOR FURTHER INVESTIGATION:
A series circuit has the same current throughout the circuit. Find this current for each setting of R_2 by applying Ohm's law to R_1 by dividing the voltage across R_1 by the resistance of R_1. Then, on Plot 5–2, graph the current as a function of the resistance of R_2. On the same graph, plot the voltage measured across R_2 as a function of the resistance of R_2. Where is the product of the current times the voltage a maximum?

APPLICATION PROBLEM:
Figure 5–3 shows a variation of the circuit in this experiment in which a light-emitting diode (LED) is used in place of R_2. The LED is a polarized component and must be put in the circuit in the correct direction. Design a measurement procedure that will enable you to determine the power in the LED as the power supply is varied. Start V_S at 1.0 V and increase it in 1.0 V increments to +12 V. Graph the power in the LED as a function of the power supply voltage in Plot 5–3. What can you conclude from this circuit?

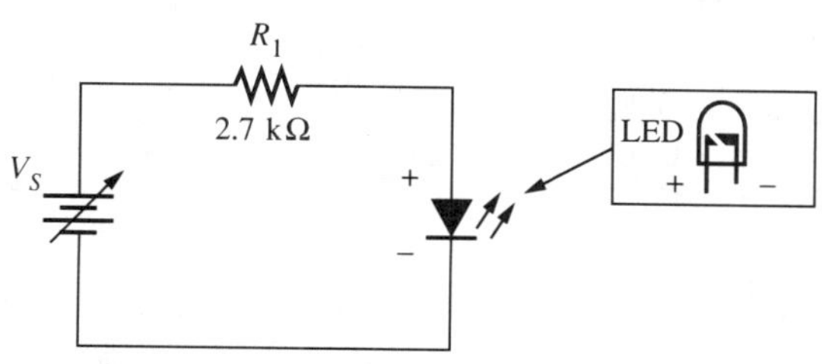

Figure 5–3

Report for Experiment 5

Name ____________
Date ____________
Class ____________

OBJECTIVES:

PROCEDURE:

DATA:

Table 5–1

Component	Listed Value	Measured Value
R_1	2.7 kΩ	

Table 5–2

Measured value of		
Resistance	Voltage	Current
kΩ	V	mA
Computed Power		
$P = IV$	$P = I^2R$	$P = \frac{V^2}{R}$
mW	mW	mW

Table 5–3

Variable Resistance Setting (R_2)	V_1 (measured)	V_2 (measured)	Power in R_2 $P_2 = \frac{V_2^2}{R_2}$
0.5 kΩ			
1.0 kΩ			
2.0 kΩ			
3.0 kΩ			
4.0 kΩ			
5.0 kΩ			
7.5 kΩ			
10.0 kΩ			

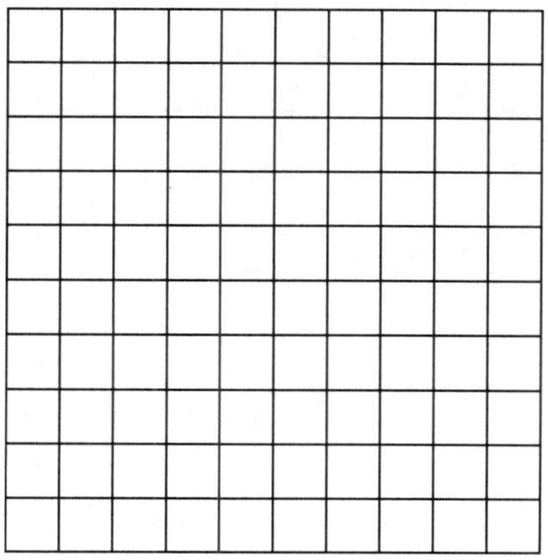

Plot 5–1

RESULTS AND CONCLUSION:

FURTHER INVESTIGATION RESULTS:

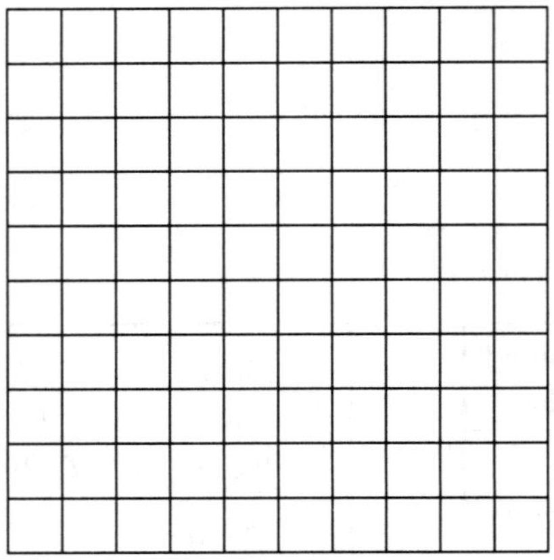

Plot 5–2

APPLICATION PROBLEM RESULTS:

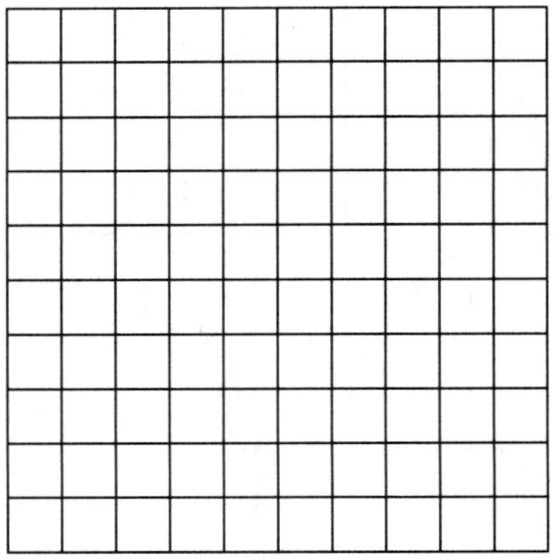

Plot 5–3

EVALUATION AND REVIEW QUESTIONS:

1. In the first part of the experiment, you computed the power in a resistor using three different equations. Why might the results in each case be slightly different?

2. For the circuit in Figure 5–1, assume a student accidentally set the power supply to 24 V instead of 12 V.
   (a) How much power would be dissipated in the resistor? ____________
   (b) Would a 1/4 W resistor be adequate for this case? ____________

3. For the circuit in Figure 5–2, what was happening to the *total power* in the circuit as the resistance of R_2 was increasing? Explain your answer.

4. A 1.5 kΩ resistor is found to have 22.5 V across it.
   (a) What is the current in the resistor? ____________
   (b) What is the power dissipated in the resistor? ____________
   (c) Could a 1/4 W resistor be used in this application? Explain your answer.

5. What physical characteristic determines the power rating of a resistor?

6. What is the smallest value of resistance that can be used across 10 V if the power dissipated is not to exceed 0.5 W?

TECH TIP REVIEW:

Each of the following questions relate to the TECH TIP circuit shown in Figure 4–12 of the Floyd text. Having constructed the resistor box, you are asked questions about it; can you provide the answers?

1. Although the resistor box is rated for 4 V, an experimenter wants to use the 1 kΩ and larger resistors across a 25 V source and asks if it is okay. Your reply is:

2. A number of batteries are to be tested using the resistor box to verify that each one will provide approximately 1 W of power. The battery voltages are between 3.0 and 3.2 V. You are asked which resistor to use for the test and how to tell if the battery is really producing the required power. Your reply is:

6 Series Circuits

OBJECTIVES:

After performing this experiment, you will be able to:

1. Use Ohm's law to find the current and voltages in a series circuit.
2. Apply Kirchhoff's voltage law to a series circuit.

READING:

Floyd, *Principles of Electric Circuits,* Sections 5–1 through 5–6

MATERIALS NEEDED:

Resistors:

One 330 Ω, one 1.0 kΩ, one 1.5 kΩ, one 2.2 kΩ

One dc ammeter, 0–10 mA

SUMMARY OF THEORY:

The current in a resistor is directly proportional to the voltage across the resistor as stated by Ohm's law. Consider the simple circuit illustrated in Figure 6–1. The source voltage is the total current multiplied by the total resistance. This can be stated in equation form as

$$V_S = I_T R_T$$

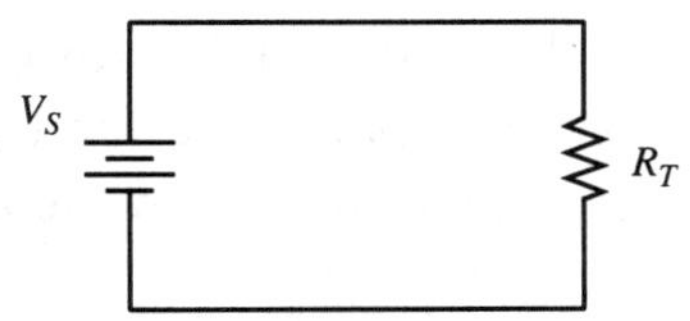

Figure 6–1

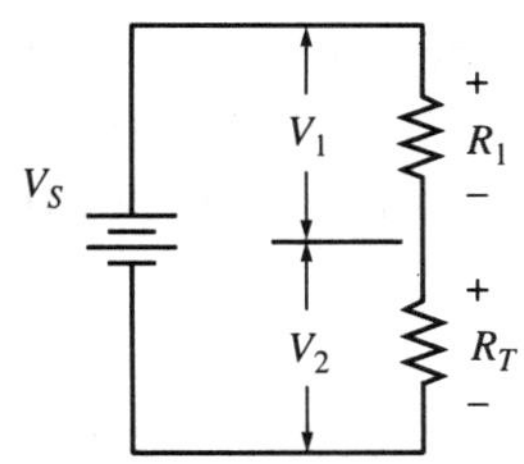

Figure 6–2

In a series circuit, the circuit elements are connected together with only one path for current. Figure 6–2 illustrates a series circuit with two resistors. When we join resistors in series, the total resistance is the sum of the individual resistors. The total resistance for the Figure 6–2 is

$$R_T = R_1 + R_2$$

Substituting this equation into Ohm's law gives

$$V_S = I_T(R_1 + R_2)$$

Multiplying both terms by I_T results in

$$V_S = I_T R_1 + I_T R_2$$

The identical current, I_T, must flow through each resistor. This causes a voltage drop across each resistor, which can be expressed as

$$V_S = V_1 + V_2$$

This result illustrates that the source voltage is equal to the sum of the voltage drops across the resistors. This relationship is called *Kirchhoff's voltage law,* which is more precisely stated as

> The algebraic sum of all voltage rises and drops around any single closed path in a circuit is equal to zero.

It is important to pay attention to the polarity of the voltages. Current from the source creates a voltage drop across the load. The voltage drop across the load will have an opposite polarity to the source voltage, as illustrated in Figure 6–2. We may apply Kirchhoff's voltage law by using the following rules:

1. Choose an arbitrary starting point. Go either clockwise or counterclockwise from the starting point.
2. For each voltage source or load, write down the first sign you see and the magnitude of the voltage.
3. When you arrive at the starting point, equate the algebraic sum of the voltages to zero.

PROCEDURE:

1. Obtain the resistors listed in Table 6–1. Measure each resistor and record the measured value in Table 6–1. Compute the total resistance for a series connection by adding the measured values. Enter the computed total resistance in Table 6–1 in the column for the listed value.

2. Connect the resistors in series, as illustrated in Figure 6–3. Measure the total resistance of the series connection and verify that it agrees with your computed value. Enter your measured value in Table 6–1.

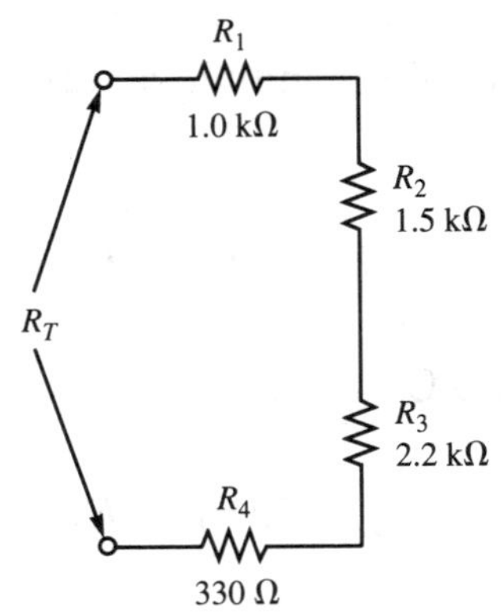

Figure 6–3

3. Complete the circuit shown in Figure 6–4, making certain that the ammeter is connected in *series;* otherwise damage to the meter may result. Before applying power, have your instructor check your circuit. Compute the current in the circuit by substituting the source voltage and the total resistance into Ohm's law; that is

$$I_T = \frac{V_S}{R_T}$$

Record the computed current in Table 6–2. Apply power, and confirm that your computed current is within experimental uncertainty of the measured current.

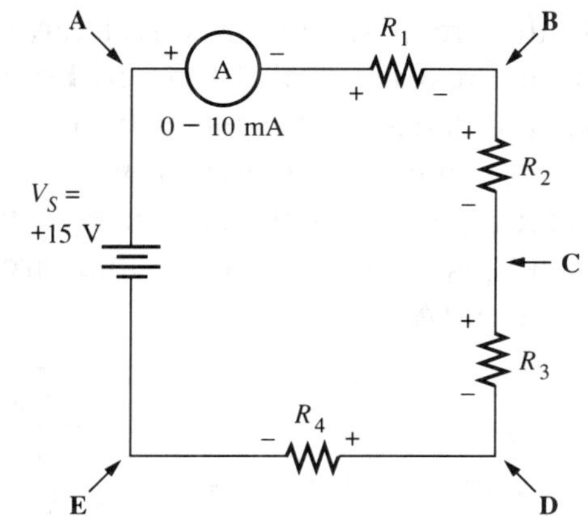

Figure 6–4

4. In a series circuit, the same current flows through all components. We can use the total current from step 3 and Ohm's law to compute the voltage drop across each resistor. Compute V_{AB} by multiplying the total current by the resistance between **A** and **B.** Record the results as the computed voltage in Table 6–2.

5. Repeat step 4 for the other voltages listed in Table 6–2.

6. Measure and record each of the voltages listed in Table 6–2.

7. Using the source voltage (+15 V) and the *measured voltage drops* listed in Table 6–2, prove that the algebraic sum of the voltage rises and drops is zero (within experimental uncertainty). Do this by applying the rules listed in the Summary of Theory. Write the algebraic sum of the voltages on the first line of Table 6–3. The polarities of voltages are shown on Figure 6–4.

8. Repeat step 7 by starting at a different point in the circuit and traversing the circuit in the opposite direction. Write the algebraic sum of the voltages on the second line of Table 6–3.

9. Open the circuit at point **B.** Measure the voltage across the open circuit. Call this voltage V_{open}. Prove that Kirchhoff's voltage law is still valid for the open circuit. Write the algebraic sum of the voltages on the third line of Table 6–3.

FOR FURTHER INVESTIGATION:

Resistors R_1, R_2, and R_3 used in this experiment have the same listed values as R_1, R_2, and R_3 from Experiment 4. Refer to your results of the current-voltage curve on Plot 4–1. Using the measured voltage in Table 6–2, find the current in the resistor on Plot 4–1. Tabulate the current read from the plot in Table 6–4. What conclusion can you make from this about the current in a series circuit?

APPLICATION PROBLEM:

A basic panel meter is a current-sensitive device. A panel meter can be converted into a current meter or voltmeter. The range of the meter is extended by limiting the current to the full-scale current for the meter. To use a panel meter as a voltmeter, a resistor is placed in series with the meter. In order to determine the required resistor, you must know the full-scale current for the meter and the internal resistance of the meter. For sensitive meters, the resistance is found indirectly to avoid damaging the meter from excessive current.

The same meter can be used as a current meter for much larger currents than allowed by the basic sensitivity of the meter by using a low-resistance "sense" resistor, R_S, as shown in the circuit of Figure 6–5. Switch S_1 is a double-pole, double-throw switch used to select between reading voltage (V) or current (A).

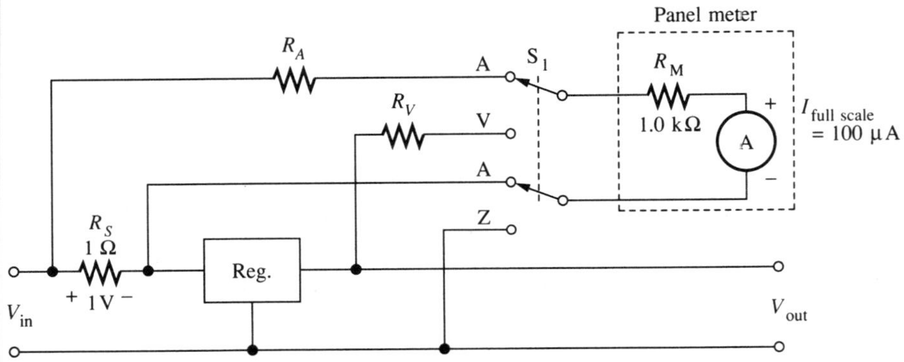

Figure 6–5

For the purpose of this application problem, assume that you have a meter with a full-scale current of 100 μA and an internal resistance of 1.0 kΩ. The meter is to be used on a regulated power supply in which, in the voltage position, the maximum reading (at full scale) is to be 15 V. In the current position, the meter will read a maximum current of 1 A. Note that 1 A in R_S produces 1 V, which, in turn, must cause only 100 μA to flow in the meter.

Determine the values of R_A and R_V in Figure 6–5 to give the required 100 μA in the meter for the full-scale voltage of 15 V and current of 1 A. Then devise an experimental check of your solution to this problem. It is not necessary to use a panel meter in your experimental check, and the sense resistor, R_S, can be replaced with a 1 V source to simulate the maximum current of 1 A.

Report for Experiment 6

Name ____________________
Date ____________________
Class ____________________

OBJECTIVES:

PROCEDURE:

DATA:

Table 6–1

Component	Listed Value	Measured Value
R_1	1.0 kΩ	
R_2	1.5 kΩ	
R_3	2.2 kΩ	
R_4	330 Ω	
R_T		

Table 6–2

	Computed Value	Measured Value
I_T		
V_{AB}		
V_{BC}		
V_{CD}		
V_{DE}		

Table 6–3

Step Number	Kirchhoff's Voltage Law (Measured Values)
7	
8	
9	

RESULTS AND CONCLUSION:

FURTHER INVESTIGATION RESULTS:

Table 6–4

	Current Read From Plot 4–1
I_1	
I_2	
I_3	

Conclusion from these data: ______________________________

APPLICATION PROBLEM RESULTS:

EVALUATION AND REVIEW QUESTIONS:

1. Why doesn't the starting point for summing the voltages around a closed loop make any difference?

2. Kirchhoff's voltage law applies to any closed path, even one without current. How did the result of step 9 show that this is true?

3. Based on the result you observed in step 9, what voltage would you expect in a 110 V circuit across an open (blown) fuse?

4. Use Kirchhoff's voltage law to find V_X in Figure 6–6.

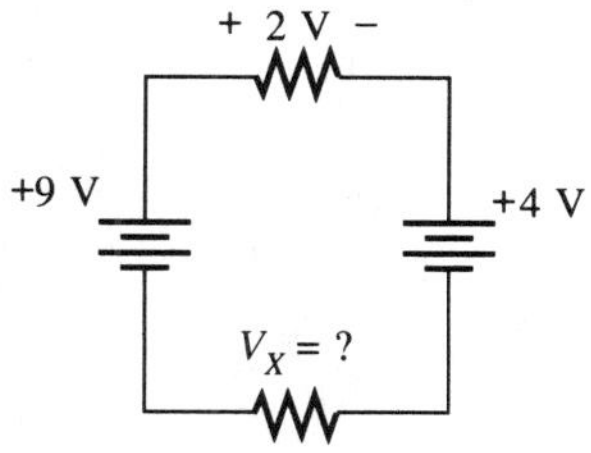

Figure 6–6

5. A 10 Ω resistor is in series with a bulb and a 12 V source.
   (a) If 8 V is across the bulb, what voltage is across the resistor? __________
   (b) What is the current in the circuit? __________
   (c) What is the resistance of the bulb? __________

6. A student wishes to limit the current to an LED (light-emitting diode) to 10.0 mA. The source voltage is + 5 V and the diode drops 1.8 V.
   (a) What value resistance is required? __________
   (b) What power is dissipated in the resistor? __________

7 The Voltage Divider

PSpice

OBJECTIVES:

After performing this experiment, you will be able to:

1. Apply the voltage divider rule to series resistive circuits.
2. Design a voltage divider to meet a specific voltage output.
3. Confirm experimentally the circuit designed in step 2.
4. Determine the range of voltages available when a variable resistor is used in a voltage divider.

READING:

Floyd, *Principles of Electric Circuits,* Sections 5–7 through 5–12

MATERIALS NEEDED:

Resistors:

One 330 Ω, one 470 Ω, one 680 Ω, one 1.0 kΩ

One 1 kΩ potentiometer

For Further Investigation: One 1.0 kΩ resistor, one 10 kΩ resistor, one 100 kΩ resistor

Application Problem: Resistors to be determined by student

SUMMARY OF THEORY:

A voltage divider consists of two or more resistors connected in series with a voltage source. Voltage dividers are used to obtain a specific voltage from a larger source voltage. The application of voltage dividers is also frequently useful for more complex circuit analysis. The application of Ohm's law to a series circuit leads us directly to the equation for a voltage divider. Consider the series circuit illustrated in Figure 7–1. The current in the circuit, I_T, and the output voltage, V_X, can be written as

$$I_T = \frac{V_S}{R_T}$$

$$V_X = I_T R_2$$

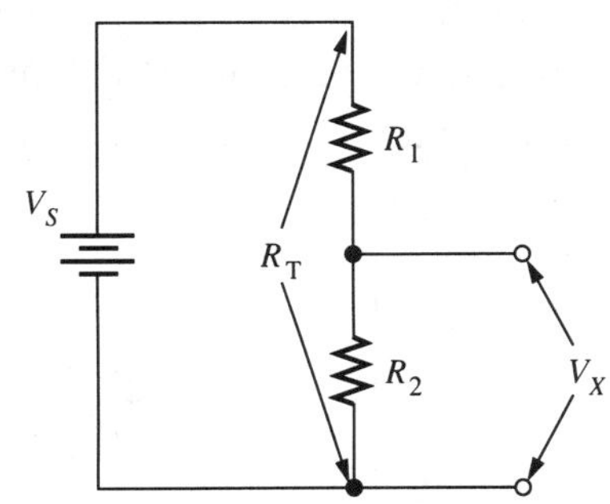

Figure 7–1

Substituting the first equation into the second gives the voltage divider equation for the two-resistor divider, as follows:

$$V_X = I_T R_2 = \left(\frac{V_S}{R_T}\right) R_2 = V_S\left(\frac{R_2}{R_T}\right)$$

This result shows the basic idea behind a voltage divider—that the input voltage across series resistors is divided in direct proportion to the resistance. A larger resistor will have a larger fraction of the input voltage across it and a smaller resistor will have a smaller fraction of the input voltage across it. That's why it's called a voltage divider. The voltage divider formula can be extended to find the voltage in a series circuit for any number of resistors. It can be written:

$$V_X = V_S\left(\frac{R_X}{R_T}\right)$$

where R_X represents the resistance between the output terminals.

The last equation is a more general form of the voltage divider equation. It can be stated as: The output voltage from a voltage divider is equal to the input voltage multiplied by the ratio of the resistance between the output terminals to the total resistance. When several resistors are used, the output is generally taken with respect to the ground reference for the divider, as shown in Figure 7–2. In this case the output voltage can be found by substituting the value of R_2 and R_3 for R_X, as shown in Figure 7–2.

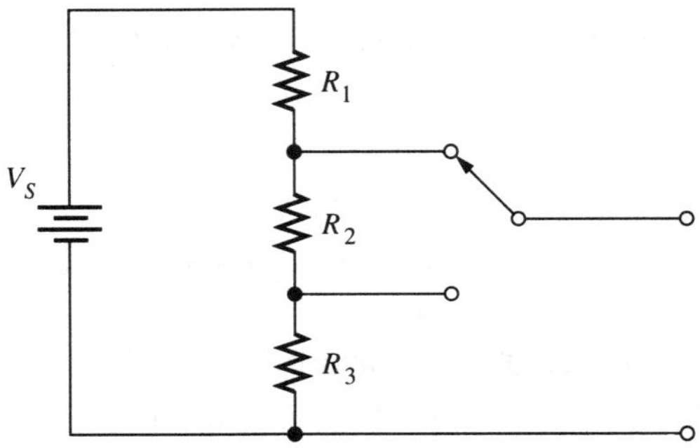

$$V_X = V_S\left(\frac{R_2 + R_3}{R_1 + R_2 + R_3}\right)$$

$$V_X = V_S\left(\frac{R_X}{R_T}\right)$$

Figure 7–2

Voltage dividers can be made to obtain variable voltages by using a potentiometer. The full range of the input voltage is available at the output, as illustrated in Figure 7–3(a). If one desires to limit the output voltage, this can be done by using fixed resistors in series, as illustrated in Figure 7–3(b).

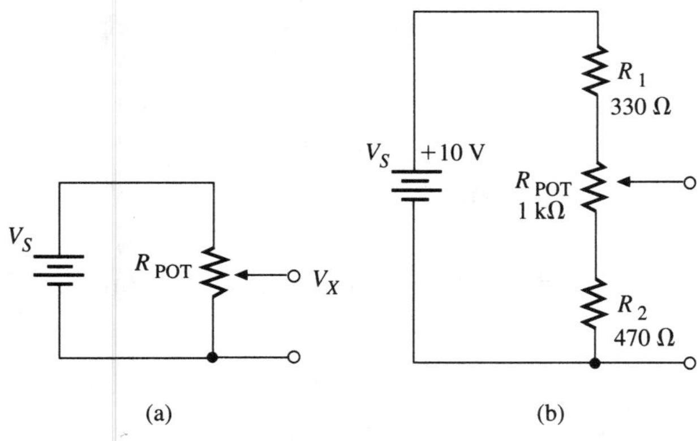

Figure 7–3

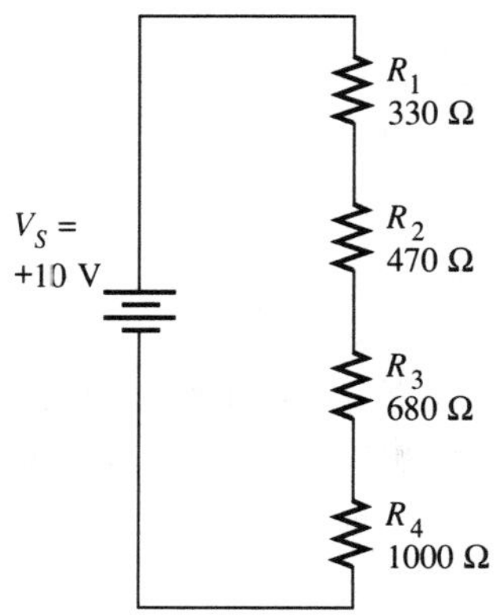

Figure 7–4

PROCEDURE:

1. Obtain the resistors listed in Table 7–1. Measure each resistor and record the measured value in Table 7–1, column 3. Compute the total resistance for a series connection by adding the measured values. Enter the computed total resistance in Table 7–1.

2. Connect the resistors in the series circuit illustrated in Figure 7–4. With the power off, measure the total resistance of the series connection and verify that it agrees with your computed value.

3. Apply the voltage divider rule to each resistor, one at a time, to compute the expected voltage across that resistor. Use the measured values of resistance and a source voltage of +10 V. Record the computed voltages (V_X) in Table 7–1, column 4.

4. Turn on the power and measure the voltage across each resistor. Record the measured voltage drops in Table 7–1, column 5. Your measured voltages should agree with your computed values.

5. Observe the voltages measured in step 4. In the space provided in the laboratory report, draw the same voltage divider as in Figure 7–4, but show on your drawing the output connection that will produce a voltage of approximately 6.8 V.

6. Assume you needed a voltage divider with a +5.0 V output from a source voltage of +10 V. (This implies that you need to take the output voltage across one-half the total resistance.) Design a new voltage divider using just three of the fixed resistors in this experiment: R_1, R_3, and R_4. Draw your design in the space provided in the laboratory report.

7. Construct the circuit you designed in step 6 and measure the output voltage. Indicate the measured output voltage on your drawing.

8. Design another voltage divider that provides a +7.5 V output from a +10 V input. Choose *only two* resistors from this experiment for your design. Draw your circuit in the space provided in the report. Then, construct the circuit and measure the actual voltage of the output. Indicate the measured output voltage on your drawing.

9. The circuit shown in Figure 7–3(b) uses a 1.0 kΩ potentiometer and R_1 and R_2 to limit the range of voltages. Assume V_S is +10 V. Use the voltage divider formula to compute the minimum and maximum voltages available from this circuit. Record your computed values in Table 7–2.

10. Construct the circuit computed in step 9. Measure the minimum and maximum output voltages. Record the measured minimum and maximum output voltages in Table 7–2.

FOR FURTHER INVESTIGATION:

The voltage dividers in this experiment were *unloaded*—that is, they were not required to furnish current to a load. If a load is put on the output, then current is supplied to the load and the output voltage of the divider changes. Investigate this effect by placing the three load resistors listed in Table 7–3 on the voltage divider from this experiment as illustrated in Figure 7–5. Summarize your findings about the effect of the load resistors on the voltage divider output.

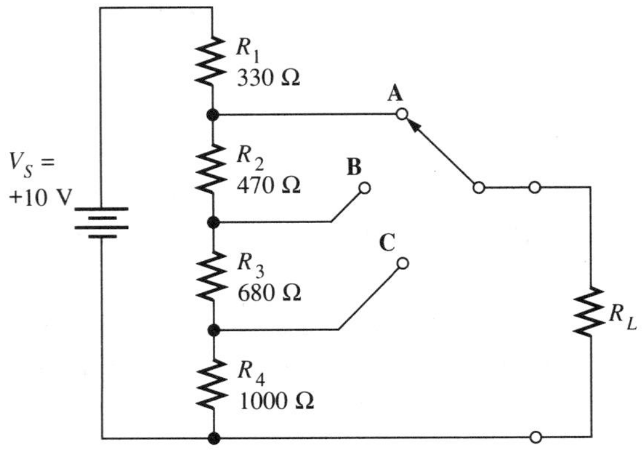

Figure 7–5

APPLICATION PROBLEM:

Voltage dividers are widely used in electronics. One application is shown in the regulated power supply circuit in Figure 7–6. A fraction of the regulated output voltage is returned to the comparator by the voltage divider. The comparator, in turn, compares the voltage divider output with a constant voltage source and controls the series regulator. The requirements for the voltage divider are as follows:

1. There should be a minimum of 3 mA of current in the divider string. There is no current required by the comparator.
2. The regulated output voltage depends on the position of the variable-tap on R_1. When R_1 is at the highest position (see Figure 7–6(b)), the regulated output voltage and the voltage divider output are both 10.0 V.
3. When R_1 is at the lowest position (see Figure 7–6(c)), the regulated output voltage rises to 15.0 V and the voltage divider output is still 10.0 V.

Compute the resistors that will satisfy the design requirements. Test your circuit using a variable power supply to simulate the output of the series regulator and show your results in the Application Problem part of your report.

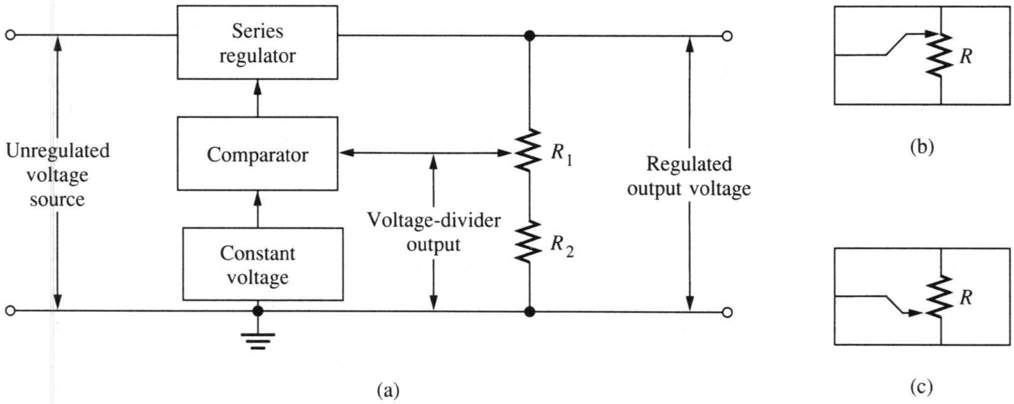

Figure 7–6

PSpice EXAMPLE:

Overview:

PSpice is a menu driven simulation program that allows you to design and simulate circuits on a computer. Pspice is a modified version of the original SPICE program; PSpice is designed to be used on personal computers. PSpice from MicroSim Corporation is the most popular of several versions on the market and was used to test the example programs in this manual.

PSpice is based on nodal analysis of a circuit. The user starts by creating a source file for the circuit that describes the circuit and the analysis desired. This file is entered into the computer and run under PSpice, creating an output file. The output file can be printed or plotted or analyzed further by a graphics postprocessor called PROBE.

The user's source file can be broken into three basic statement types. These are *data statements,* used to define the circuit, *control statements* that describe the analysis desired, and *output statements* that specify the voltages and currents to be printed. Floyd's text introduces the creation of source files in a step by step process as you are ready. The purpose of the programs in the lab manual is to give you additional examples from your experience, that you may want to try on PSpice (a few new statements are used that will be introduced later). The examples given are all from circuits shown in the laboratory exercises.

To enter a source file, start PSpice, enter the Files menu and select Current File, then enter a filename. From the Files menu, select Edit. This brings up the PSpice editor. Enter your source file (an example follows). Press the Esc key to exit the editor. From the Analysis menu, select Run PSpice. After running the program, you can look at the output by entering the Files menu and selecting the Browse Output option. A hard copy of the output file can be sent to the printer.

The following source file is from Figure 7–3. Node 4 is the output node. The output will show you how the voltage of node 4 changes as the potentiometer is moved from minimum to maximum resistance in 10 steps.

```
LAB7 FIG7-3
VS 3 0 10
R1 3 2 330
.PARAM SET=. 1 R=1K
STEP PARAM(SET) .1, 1, .1
RTOP 2 4 {(1.001-SET)*R}
RBOT 4 1 {(.001+SET)*R}
R2 1 0 470
. OPT NOPAGE
. END
```

Report for Experiment 7

Name ____________________
Date ____________________
Class ____________________

OBJECTIVES:

PROCEDURE:

DATA:

Table 7–1

Resistor	Listed Value	Measured Value	$V_x = V_s\left(\frac{R_x}{R_T}\right)$	V_x (measured)
R_1	330 Ω			
R_2	470 Ω			
R_3	680 Ω			
R_4	1000 Ω			
Total			10.0 V	

(Use all 4 resistors.)

Circuit for Step 5

(Use 3 resistors.)

Circuit for Step 6

(Use 2 resistors.)

Circuit for Step 8

Table 7–2

	Computed	Measured
V_{MIN}		
V_{MAX}		

RESULTS AND CONCLUSION:

FURTHER INVESTIGATION RESULTS:

(a) Data:

Table 7–3

Switch Position	Measured Voltages		
	R_L = 1.0 kΩ	R_L = 10 kΩ	R_L = 100 kΩ
A			
B			
C			

(b) Results:

APPLICATION PROBLEM RESULTS:

EVALUATION AND REVIEW QUESTIONS:

1. (a) If all of the resistors in Figure 7–4 were 10 times larger than the specified values, what would happen to the output voltage?

   (b) What would happen to the power dissipated in the voltage divider?

2. Refer to Figure 7–3(b). Assume V_s is +10.0 V.
   (a) If R_1 is open, what is the output voltage? ____________
   (b) If R_2 is open, what is the output voltage? ____________

3. If a student used a potentiometer in the circuit of Figure 7–3(b) that was 10.0 kΩ instead of 1.0 kΩ, what would happen to the *range* of output voltages?

4. For the circuit in Figure 7–7, compute the output voltage for each position of the switch.

   V_A = ____________
   V_B = ____________
   V_C = ____________
   V_D = ____________

5. Compute the minimum and maximum voltage available from the circuit shown in Figure 7–8.

   V_{MIN} = ____________ V_{MAX} = ____________

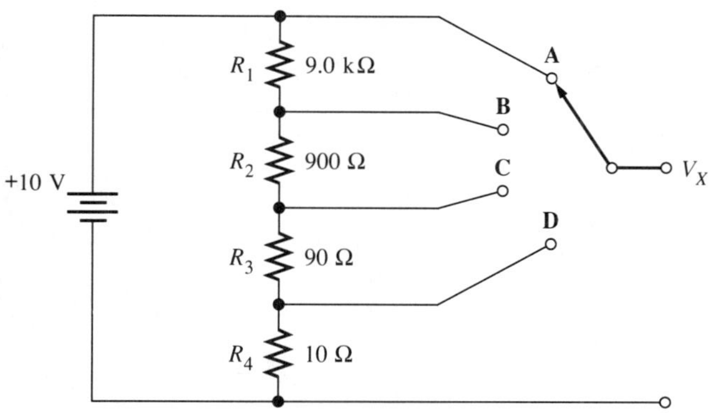

Figure 7–7

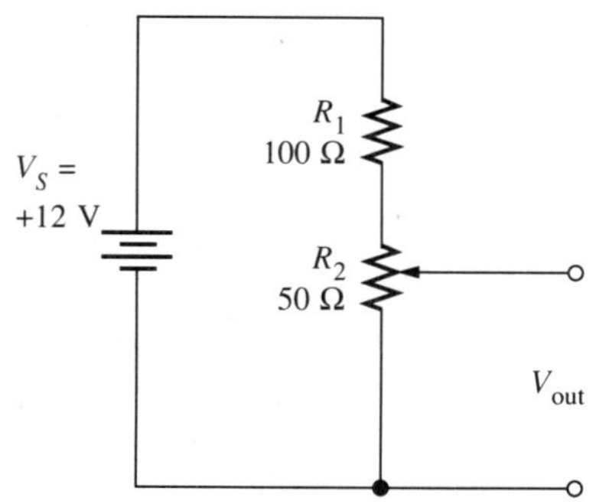

Figure 7–8

6. (a) Compute the power dissipated by each resistor in Figure 7–8.

(b) Why isn't the power dissipated affected by V_{out}?

8 Circuit Ground

OBJECTIVES:

After performing this experiment, you will be able to:

1. Use voltages measured with respect to ground to compute the voltage drop across a resistor.
2. Explain the meaning of circuit ground and subscripts used in voltage definitions.

READING:

Floyd, *Principles of Electric Circuits,* Sections 2–6 and 5–9 through 5–11

MATERIALS NEEDED:

Resistors:

One 330 Ω, one 680 Ω, one 1.0 kΩ

For Further Investigation: Second +5 V power supply with "floating" common connection

Application Problem: One additional 1.0 kΩ resistor.

SUMMARY OF THEORY:

Energy is required to move a charge from a point of lower potential to one of higher potential. Voltage is a measure of this energy per charge. Energy is given up when a charge moves from a point of higher potential to one of lower potential.

Voltage is always measured with respect to some point in the circuit. For this reason, only potential *differences* have meaning. We can refer to the voltage *across* a component in which case the reference is one side of the component. Alternatively, we can refer to the voltage at some point in the circuit. In this case the reference point is assumed to be "ground." Circuit ground is usually called *reference ground* to differentiate it from the potential of the earth, which is called *earth ground.* Circuit or earth grounds are shown with the symbol used on Figure 8–1.

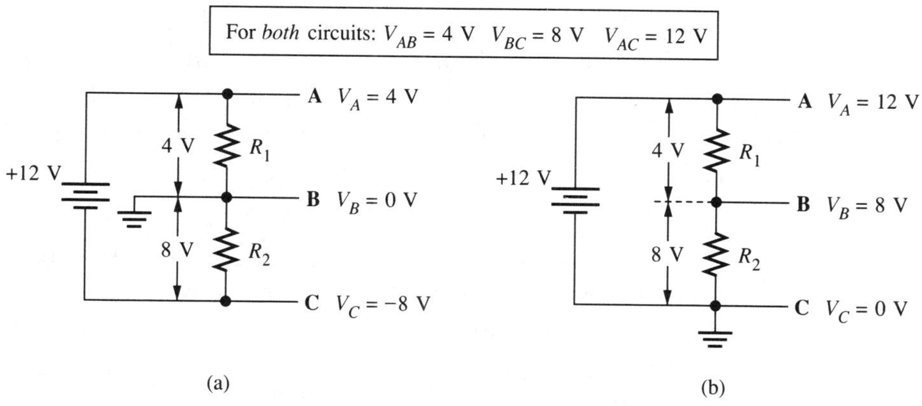

Figure 8–1

An analogy can clarify the meaning of reference ground. Assume a building has two floors below ground level. The floors in the building could be numbered from the ground floor by numbering the lower floors with negative numbers. Alternatively, the reference for numbering the floors could be made the lowest floor in the basement. Then all floors would have a positive floor number. The choice of the numbering system does not change the height of the building, but it does change each floor number. Likewise, the ground reference is used in circuits as a point of reference for voltage measurements. The circuit is not changed by the ground reference chosen.

Figure 8–1 illustrates the same circuit with two different ground reference points. The circuit in Figure 8–1(a) has as its reference point **B.** Positive and negative voltages are shown. If the reference point is moved to point **C,** the circuit voltages are all positive. Voltage is always measured between two points. To define the two points, subscripts are used. The voltage difference (or simply voltage) between points **A** and **B** is written as V_{AB}. If a single subscripted letter is shown, the voltage is defined between the lettered point and the circuit's reference ground.

PROCEDURE:

1. Measure three resistors with the listed values given in Table 8–1. Record the measured values in Table 8–1.

2. Construct the circuit shown in Figure 8–2. Set the power supply to +10 V. Measure the voltage *across* each resistor in the circuit. For example, to measure the voltage across R_1, place one probe of your DMM on point **A** and the other probe on point **B.** Enter the measured voltages in Table 8–2.

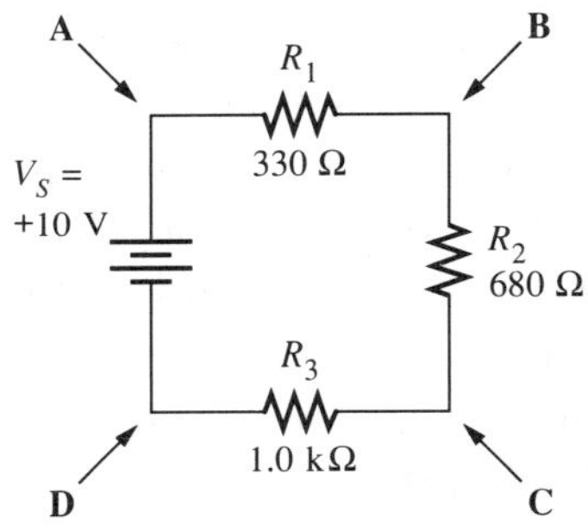

Figure 8–2

3. Assign point **D** as the reference ground for the circuit. Measure the voltage at points **A, B,** and **C** with respect to point **D.** The voltage readings are made with the reference probe of your voltmeter connected to point **D.** Enter the measured values in Table 8–3. Then use the measured voltages to compute the voltage differences V_{AB}, V_{BC}, and V_{CD}.

4. Now measure the voltages in the circuit with respect to point **C.** The circuit is *not changed.* Only the reference point changes. Move the reference probe of your voltmeter to point **C.** This point will now represent ground. The voltage at point **D** now has a negative value. Enter the measured voltages in Table 8–4. Compute the voltage differences as before and enter them in Table 8–4.

5. Move the circuit reference point to point **B.** Again, there is no change to the circuit other than the reference ground. Repeat the measurements of the voltages with respect to circuit ground. Compute the voltage differences and enter the data in Table 8–5.

6. Now make point **A** the reference point and repeat the measurements. Enter the data in Table 8–6.

FOR FURTHER INVESTIGATION:

Replace the +10 V supply used in this experiment with two +5 V supplies in series. Attach the +5 V output of one supply to the common of the second supply. Call this point the reference ground for the circuit. Measure the voltages throughout the circuit. Summarize your results.

APPLICATION PROBLEM:

Sometimes, in digital logic circuits, it is necessary to compare an unknown voltage level with a specified threshold level to see if the unknown voltage is above or below the specification. A series voltage divider circuit can be used to provide both positive and negative reference voltages for logic circuits. The circuit shown in Figure 8–3 is needed to provide input logic threshold voltages for both TTL (transistor-transistor-logic) and ECL (emitter-coupled-logic). TTL uses positive voltages, whereas ECL uses negative voltages. The voltages required are shown on Figure 8–3 and can be obtained from a single adjustable power supply. Use the resistors from this experiment plus one additional 1.0 kΩ resistor to design a voltage divider that produces the voltages shown. Each voltage should be within 5% of the required voltage. Set up your circuit, measure the voltages with respect to ground, and report on your results.

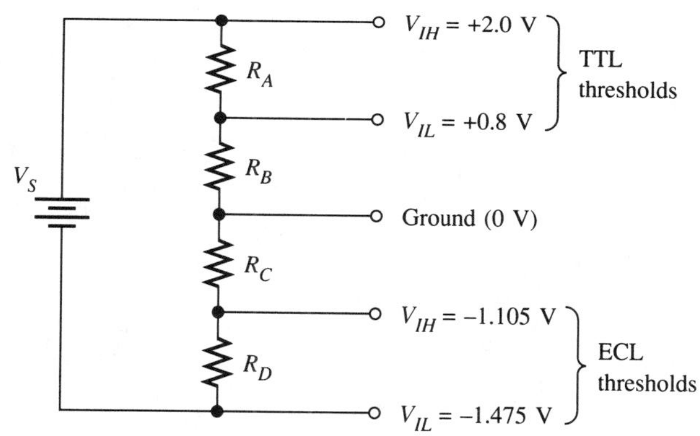

Figure 8–3

Report for Experiment 8

Name ______________
Date ______________
Class ______________

OBJECTIVES:

PROCEDURE:

DATA:

Table 8–1

Component	Listed Value	Measured Value
R_1	330 Ω	
R_2	680 Ω	
R_3	1.0 kΩ	

Table 8–2

	Measured Value
V_S	
V_{AB}	
V_{BC}	
V_{CD}	

Table 8–3

	Measured Voltage
V_A	
V_B	
V_C	
V_D	0.0 V (ref)

Voltage Difference Calculation
$V_{AB} = V_A - V_B =$
$V_{BC} = V_B - V_C =$
$V_{CD} = V_C - V_D =$

Table 8–4

	Measured Voltage
V_A	
V_B	
V_C	0.0 V (ref)
V_D	

Voltage Difference Calculation
$V_{AB} = V_A - V_B =$
$V_{BC} = V_B - V_C =$
$V_{CD} = V_C - V_D =$

Table 8–5

	Measured Voltage
V_A	
V_B	0.0 V (ref)
V_C	
V_D	

Voltage Difference Calculation
$V_{AB} = V_A - V_B =$
$V_{BC} = V_B - V_C =$
$V_{CD} = V_C - V_D =$

Table 8–6

	Measured Voltage
V_A	0.0 V (ref)
V_B	
V_C	
V_D	

Voltage Difference Calculation
$V_{AB} = V_A - V_B =$
$V_{BC} = V_B - V_C =$
$V_{CD} = V_C - V_D =$

RESULTS AND CONCLUSION:

FURTHER INVESTIGATION RESULTS:

APPLICATION PROBLEM RESULTS:

EVALUATION AND REVIEW QUESTIONS:

1. Compare the *Voltage Difference Calculation* in Tables 8–3 through Table 8–6. Does the circuit's reference point have any effect on the voltage differences across any of the resistors? Explain your answer.

2. Define the term *reference ground.*

3. If you measured V_{AB} as +12.0 V, what is V_{BA}?

4. Assume $V_M = -220$ V and $V_N = -150$ V. What is V_{MN}?

5. If a test point in a circuit is marked +5.0 V and a second test point is marked −3.3 V, what voltage reading would you expect on a voltmeter connected between the two test points? Assume the reference lead on the meter is at the lowest potential.

6. Figure 8–4 shows a field-effect transistor with certain dc voltages given.
   (a) What is V_{GS}? ____________
   (b) What is V_{DS}? ____________

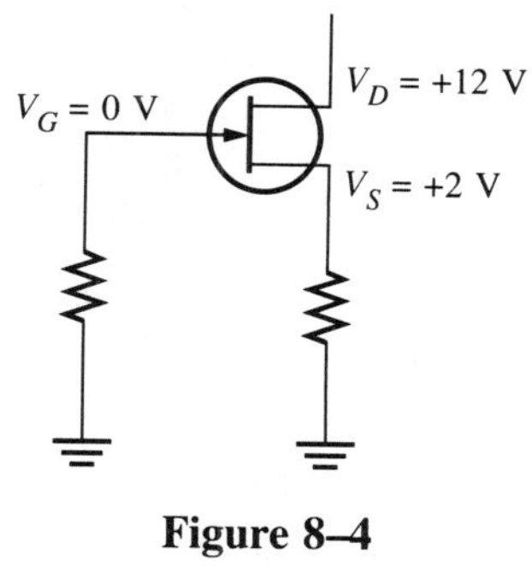

Figure 8–4

TECH TIP REVIEW:

Each of the following problems relate to the TECH TIP voltage-divider circuit shown in Figure 5–63 of the Floyd text. Assume a power supply is connected to the on pins 3 and 1. Determine the effect on the output for each of the following problems.

1. Assume the power supply was connected to the divider board but was set 10% low (10.8 V instead of +12 V). How would this affect the output voltages?

2. Assume all resistors were 10% larger than specified.
   (a) What effect does this have on the output voltage?

   (b) What effect does this have on the power dissipated in the divider?

3. Assume the power supply was hooked up in reverse (ground connected to pin 3; +12 V connected to pin 1). How would this error affect the output voltages?

9 Parallel Circuits

OBJECTIVES:

After performing this experiment, you will be able to:

1. Demonstrate that the total resistance in a parallel circuit decreases as resistors are added.
2. Compute and measure resistance and currents in parallel circuits.
3. Explain how to troubleshoot parallel circuits.

READING:

Floyd, *Principles of Electric Circuits,* Sections 6–1 through 6–10

MATERIALS NEEDED:

Resistors:

One 3.3 kΩ, one 4.7 kΩ, one 6.8 kΩ, one 10.0 kΩ

One dc ammeter, 0–10 mA

Application Problem: One 1.0 kΩ potentiometer

SUMMARY OF THEORY:

A *parallel* circuit is one in which there is more than one path for current to flow. Parallel circuits can be thought of as two parallel lines, representing conductors, with a voltage source and components connected between the lines. This idea is illustrated in Figure 9–1. The source voltage appears across each component. Each path for current is called a *branch*. The current in any branch is dependent only on the resistance of that branch and the source voltage.

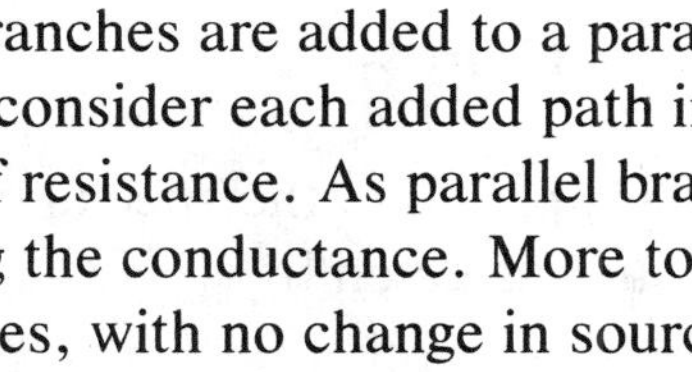

Figure 9–1

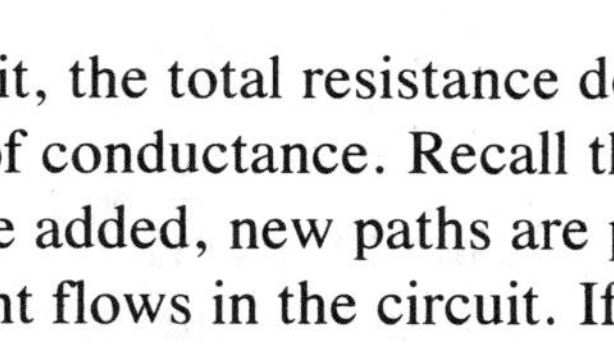

Figure 9–2

As more branches are added to a parallel circuit, the total resistance decreases. This is easy to see if you consider each added path in terms of conductance. Recall that conductance is the reciprocal of resistance. As parallel branches are added, new paths are provided for current, increasing the conductance. More total current flows in the circuit. If the total current in a circuit increases, with no change in source voltage, the total resistance must decrease according to Ohm's law. The total conductance of a parallel circuit is the sum of the individual conductances. This can be written:

$$G_T = G_1 + G_2 + G_3 + \cdots + G_i$$

By substituting the definition for resistance into the formula for conductance, the reciprocal formula for resistance in parallel circuits is obtained:

$$\frac{1}{R_T} = \frac{1}{R_1} + \frac{1}{R_2} + \frac{1}{R_3} + \cdots + \frac{1}{R_i}$$

In parallel circuits, there are junctions where two or more components are connected. Figure 9–2 shows a circuit junction labeled **A.** Since electrical charge cannot accumulate at a point, the current flowing into the junction must be equal to the current flowing from the junction. This idea is Kirchhoff's current law, which is stated as follows:

> The sum of the currents entering a circuit junction is equal to the sum of the currents leaving the junction.

One important idea can be seen by applying Kirchhoff's current law to a point next to the source voltage. The current leaving the source must be equal to the sum of the individual branch currents. Although Kirchhoff's voltage law is developed in the study of series circuits and the current law is developed in the study of parallel circuits, keep in mind that both laws are applicable to any circuit.

The voltage divider rule was developed for series circuits on the premise that the same current was in all components of a series circuit. In a parallel circuit, the same voltage is across every component. By equating the *IR* drop across each component in parallel, the current divider rule can be developed. The current divider rule has two useful forms. The first is the general equation for a parallel circuit containing several resistors, as illustrated in Figure 9–3. The general equation for the current divider is

$$I_X = \left(\frac{R_T}{R_x}\right) I_T$$

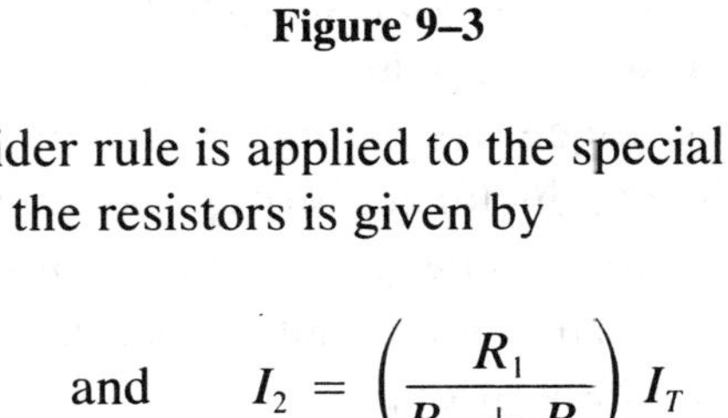

Figure 9–3

The second form of the current divider rule is applied to the special case of two resistors in parallel. The current in each of the resistors is given by

$$I_1 = \left(\frac{R_2}{R_1 + R_2}\right) I_T \quad \text{and} \quad I_2 = \left(\frac{R_1}{R_1 + R_2}\right) I_T$$

PROCEDURE:

1. Obtain the resistors listed in Table 9–1. Measure and record the value of each resistor.

2. In Table 9–2 you will tabulate the total resistance as resistors are added in parallel. (Parallel connections are indicated with two parallel lines shown between the resistors.) Enter the measured value of R_1 in the table. Then connect R_2 in parallel with R_1 and measure the total resistance, as shown in Figure 9–4. Enter the measured resistance of R_1 in parallel with R_2 in Table 9–2.

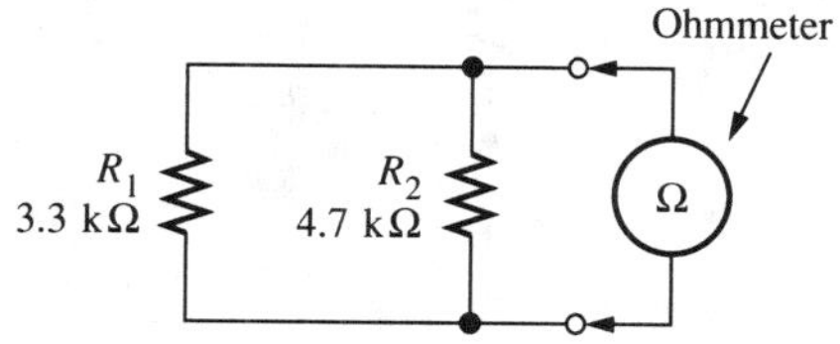

Figure 9–4

3. Add R_3 in parallel with R_1 and R_2. Measure the parallel resistance of all three resistors. Then add R_4 in parallel with the other resistors and repeat the measurement. Record your results in Table 9–2.

4. Complete the parallel circuit by adding the voltage source and the ammeter, as shown in Figure 9–5. Be certain that the ammeter is connected in series with the voltage source, as shown. If you are not sure, have your instructor check your circuit. Measure the total current and record it in Table 9–2.

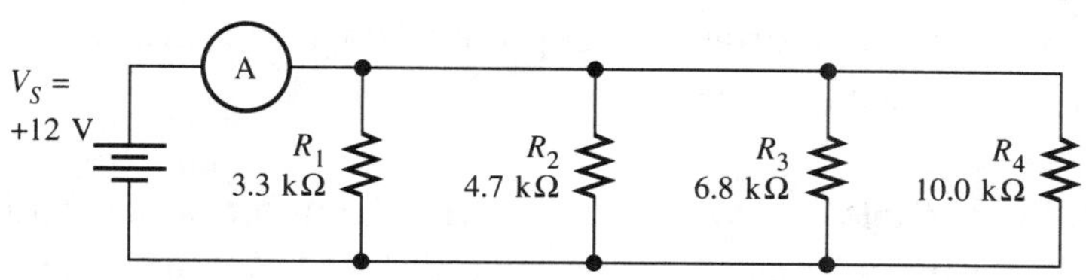

Figure 9–5

5. Measure the voltage across each resistor. If you have correctly connected them in parallel, the voltage will be the same across each resistor and equal to the source voltage.

6. Use Ohm's law to compute the branch current in each resistor. Use the source voltage and the measured resistances. Tabulate the computed currents in Table 9–3.

7. Using the currents computed in step 6, prove Kirchhoff's current law for the circuit by showing that the total current is equal to the sum of the branch currents. Write your proof in the results and conclusion section of your report.

8. Simulate a burned-out (open) resistor by removing R_1 from the circuit. Measure the new total current in the circuit. Record the current in Table 9–4.

FOR FURTHER INVESTIGATION:

Kirchhoff's current law can be applied to any junction in a circuit. The currents in this circuit were I_1, I_2, I_3, I_4, and I_T. Apply Kirchhoff's current law to these currents by writing the numerical value of the current entering and leaving each junction circled in Figure 9–6. Then verify that you computed the correct currents by measuring them with the ammeter. Summarize your results in your report.

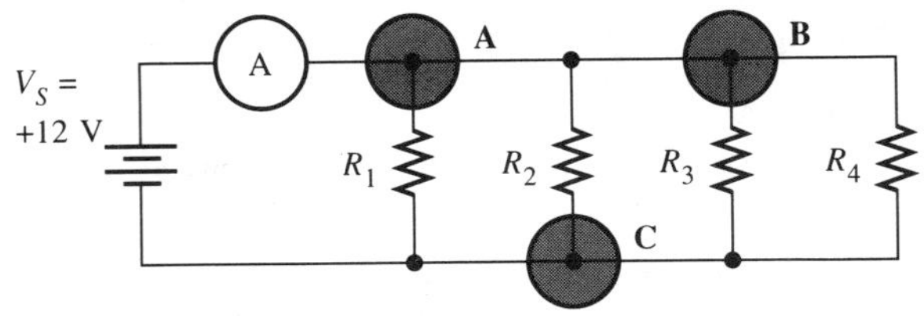

Figure 9–6

APPLICATION PROBLEM:

The range of an ammeter can be greatly increased by providing a parallel path for current around the meter. The parallel resistance is termed a *shunt*. In order to compute the proper shunt for an ammeter, the internal resistance of the meter must be known as well as its sensitivity. The sensitivity of an ammeter is the current required to produce full-scale deflection. With sensitive meters, it is best to determine the internal meter resistance indirectly to avoid damage to the meter. The following procedure is a method to determine the internal resistance of an ammeter indirectly. The method is called the full-scale deflection method:

1. Obtain a 10 kΩ resistor and place it in series with the ammeter. Connect a variable power supply as shown in Figure 9–7 and *slowly* increase the voltage until the meter is reading full scale. (*Note:* For low-resistance meters, it may be necessary to change the resistor to a lower value in order to keep the voltage at a low level.) Measure the power supply voltage at full-scale deflection.

2. Connect a 1.0 kΩ variable resistor in parallel with the meter, as shown in Figure 9–8. Adjust the variable resistor until the meter reads about 40% of full scale. (Exact reading is not critical.)

3. Set the power supply to exactly twice the voltage noted in procedure A. Then adjust the variable resistor until the meter reads exactly full scale. At this point the variable resistor has the same resistance as the internal resistance of the meter. Turn off the power supply and remove the variable resistor from the circuit. The variable resistor can now be measured with an ohmmeter.

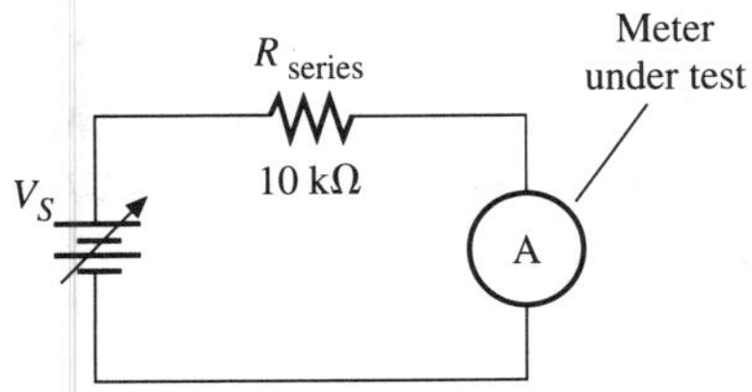

Figure 9–7

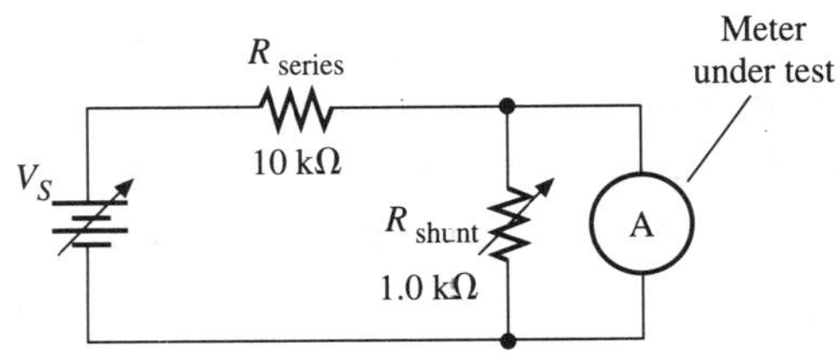

Figure 9–8

For this application problem, do the following:

1. Determine the internal resistance of the 10 mA ammeter for this experiment using the full-scale deflection method described previously. Although a 10 mA meter is not particularly sensitive and in most cases could be measured directly with an ohmmeter, this exercise will assure you understand the method.
2. Calculate the value of a shunt resistor that can be placed in parallel with your ammeter to cause it to read 50 mA full scale. The shunt will need to pass 40 mA around the meter.
3. Set a variable resistor for the value determined in step 2 and place it in parallel with the meter. The circuit is shown in Figure 9–9. Monitor the voltage across the series resistor and adjust the power supply until 5.0 V is across it. Calculate the current leaving the supply and confirm that the meter is reading one-fifth of the circuit current. Summarize your work in your report.

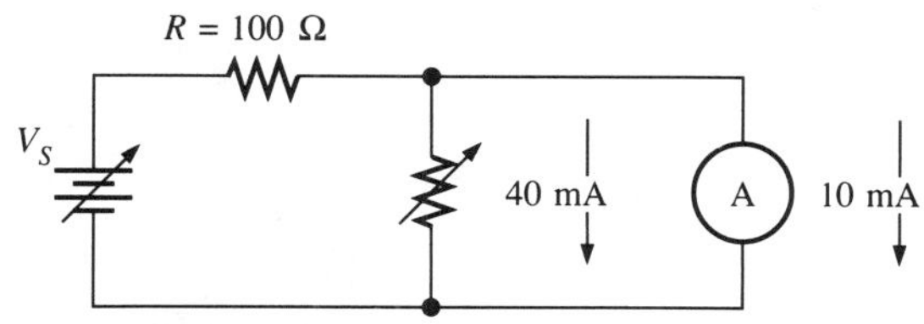

Figure 9–9

Report for Experiment 9

Name ____________
Date ____________
Class ____________

OBJECTIVES:

PROCEDURE:

DATA:

Table 9–1

Component	Listed Value	Measured Value
R_1	3.3 kΩ	
R_2	4.7 kΩ	
R_3	6.8 kΩ	
R_4	10.0 kΩ	

Table 9–2

	R_1	$R_1//R_2$	$R_1//R_2//R_3$	$R_1//R_2//R_3//R_4$
R_T (measured)				
I_T (measured)				

Table 9–3

	$I_1 = \frac{V_S}{R_1}$	$I_2 = \frac{V_S}{R_2}$	$I_3 = \frac{V_S}{R_3}$	$I_4 = \frac{V_S}{R_4}$
I (computed)				

Table 9–4

Total current with R_1 open = ________

RESULTS AND CONCLUSION:

FURTHER INVESTIGATION RESULTS:

APPLICATION PROBLEM RESULTS:

EVALUATION AND REVIEW QUESTIONS:

1. In step 8, you simulated an open resistor by removing it from the circuit. Explain how the open resistor can be found in this experiment if the only reading available is the total current.

2. If one of the resistors in this experiment were shorted, what would you expect to see happen? (Do not simulate this!)

3. Three resistors with resistances of 620 Ω, 750 Ω, and 820 Ω are in parallel across a 40 V source.

   (a) What should the total source current be?

   (b) If the measured source current was actually 118 mA, what fault could account for this?

4. The known currents for a circuit junction are shown in Figure 9–10. What is the value and direction of the unknown current, I_4?

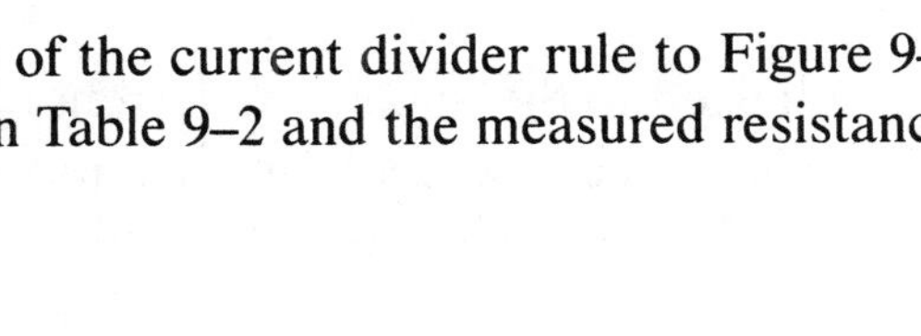

Figure 9–10

5. Show the application of the current divider rule to Figure 9–5 by using the total current and total resistance in Table 9–2 and the measured resistances from Table 9–1. Find I_1, I_2, I_3, and I_4.

6. Assume the ammeter shown in Figure 9–11 has an internal resistance of 60 Ω with a maximum (full-scale) current of 25 mA. A shunt (parallel) resistance is required that will allow the meter to read a total of 250 mA by allowing 25 mA through the meter and rest through the shunt.

(a) What is the current in the shunt resistor?

(b) Specify the resistance and wattage rating of the shunt resistor.

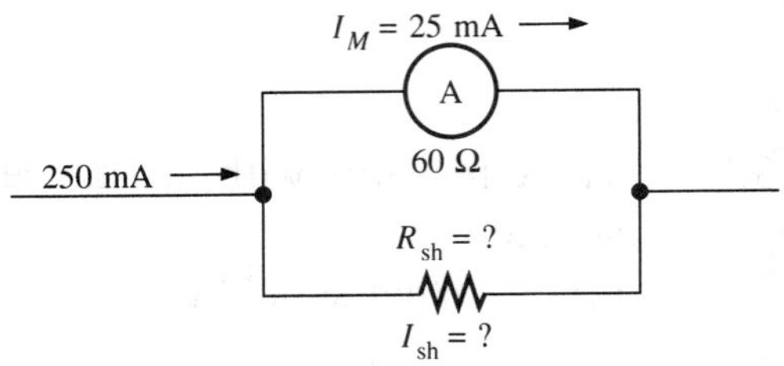

Figure 9–11

TECH TIP REVIEW:

Each of the following problems relate to the TECH TIP circuit shown in Figure 6–54 of the Floyd text. Assume you have connected the modified ammeter circuit to a meter calibrator, an instrument that can be programmed for a given output current. From the following measurements, can you determine the most probable cause for each of the following problems?

1. The meter switch is placed in the 10 mA position. The meter calibrator is programmed for 1 mA of output current but the modified ammeter reads full scale. A voltmeter check of the voltage across R_2 reveals that there are 50 mV directly across the resistor.

2. The meter calibrator is adjusted to the maximum current on each range and the modified ammeter responds correctly except for the 5 mA range. On this range, no reading is obtained on the meter, but the meter calibrator indicates that 5 mA are flowing.

10 Series-Parallel Combination Circuits

OBJECTIVES:

After performing this experiment, you will be able to:

1. Use the concept of equivalent circuits to simplify series-parallel circuit analysis.
2. Compute the currents and voltages in a series-parallel combination circuit and verify your computation with circuit measurements.

READING:

Floyd, *Principles of Electric Circuits,* Sections 7–1 through 7–8

MATERIALS NEEDED:

Resistors:

One 2.2 kΩ, one 4.7 kΩ, one 5.6 kΩ, one 10 kΩ

SUMMARY OF THEORY:

Most electronic circuits are not just series or just parallel circuits. Instead, they may contain combinations of components. Many circuits can be analyzed by applying the ideas developed for series and parallel circuits to them. Remember that in a *series* circuit the same current flows through all components and that the total resistance of series resistors is the sum of the individual resistors. By contrast, in *parallel* circuits, the applied voltage is the same across all branches and the total resistance is given by the reciprocal formula.

In this experiment, the circuit elements are connected in composite circuits containing both series and parallel combinations. The key to solving these circuits is to form equivalent circuits from the series or parallel elements. You need to recognize when circuit elements are connected in series or parallel in order to form the equivalent circuit. For example, in Figure 10–1(a) we see that the identical current must flow through both R_2 and R_3. We conclude that these resistors are in series and could be replaced by an equivalent resistor equal to their sum. Figure 10–1(b) illustrates this idea. The circuit has been simplified to an equivalent parallel circuit. After finding the currents in the equivalent circuit, the results can be applied to the original circuit to complete the solution.

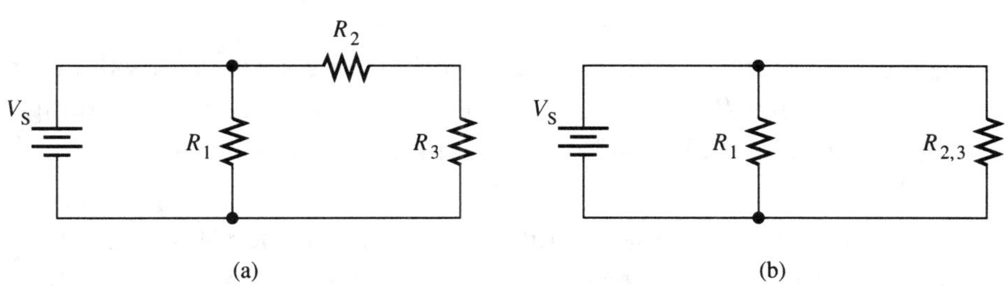

Figure 10–1

The answer to two questions will help you identify a series or parallel connection: (1) Will the *identical* current flow through both components? If the answer is yes, the components are in series. (2) Are *both ends* of one component connected directly to *both ends* of another component? If yes, the components are in parallel. The components that are in series or parallel may be replaced with an equivalent component. This process continues until the circuit is reduced to a simple series or parallel circuit. After solving the equivalent circuit, the process is reversed in order to apply the solution to the original circuit. This idea will be studied in this experiment.

PROCEDURE:

1. Measure and record the actual values of the four resistors listed in Table 10–1.

2. Connect the circuit shown in Figure 10–2. Notice that the identical current flows through R_1 and R_4, so we know that they are in series. R_2 has both ends connected directly to R_3, so these resistors are in parallel.

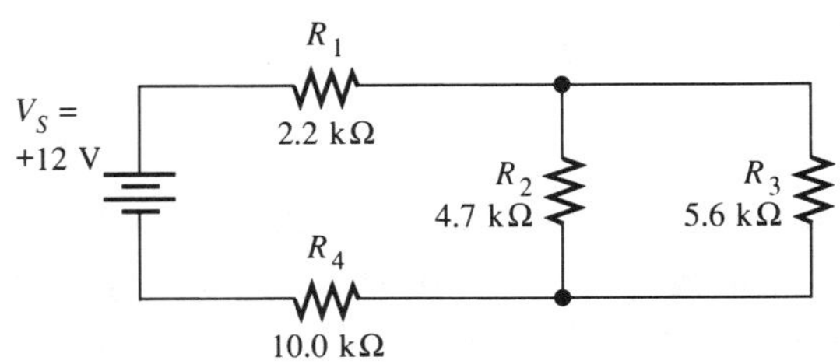

Figure 10–2

3. You can begin solving for the currents and voltages in the circuit by replacing resistors that are either in series or in parallel with an equivalent resistor. In this case, begin by replacing R_2 and R_3 with one equivalent resistor. Label the equivalent resistor $R_{2,3}$. Draw the equivalent series circuit in the space provided in the report. Show the value of all components, including $R_{2,3}$.

4. The equivalent circuit you drew in step 3 is a series circuit. Compute the total resistance of this equivalent circuit and enter it in the first two columns of Table 10–2. Then disconnect the power supply and measure the total resistance to confirm your calculation. Enter the measured total resistance, R_T, in column 3.

5. The voltage divider rule can be applied directly to the series equivalent circuit to find the voltages across R_1, $R_{2,3}$, and R_4. Find V_1, $V_{2,3}$, and V_4 using the voltage divider rule. Tabulate the results in Table 10–2 in the *Voltage Divider* column.

6. Find the total current, I_T, in the circuit by substituting the total voltage and the total resistance into Ohm's law. Enter the computed total current in Table 10–2 in the *Ohm's law* column.

7. In the equivalent series circuit, the total current flows through R_1, $R_{2,3}$, and R_4. The voltage drop across each of these resistors can be found by applying Ohm's law to each resistor. Compute V_1, $V_{2,3}$, and V_4 using this method. Enter the voltages in Table 10–2 in the *Ohm's law* column.

8. Use $V_{2,3}$ and Ohm's law to compute the current in R_2 and R_3 of the original circuit. Enter the computed current in Table 10–2. As a check, verify that the computed sum of I_2 and I_3 is equal to the computed total current.

9. Measure the voltages V_1, $V_{2,3}$, and V_4. Enter the measured values in Table 10–2.

10. Change the circuit to the circuit shown in Figure 10–3. In the space provided in your report, draw an equivalent circuit by combining the resistors that are in series. Enter the values of the equivalent resistors on your schematic drawing and in Table 10–3.

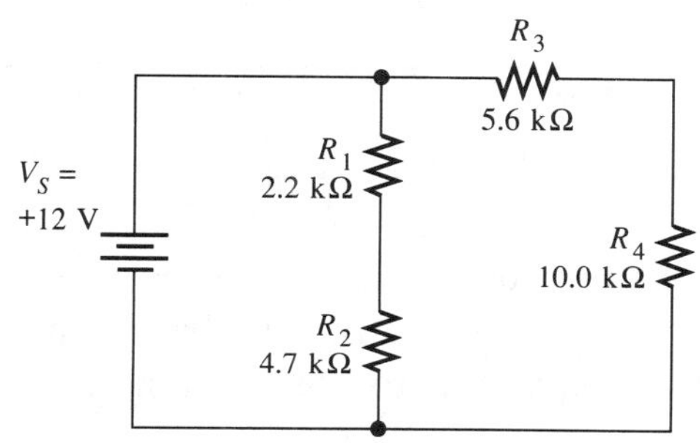

Figure 10–3

11. Compute the total resistance, R_T, of the equivalent circuit. Then apply Ohm's law to find the total current I_T. Enter the computed resistance and current in Table 10–3.

12. Complete the calculations of the circuit by solving for the remaining currents and voltages listed in Table 10–3. Then measure the voltages across each resistor to confirm your computation.

FOR FURTHER INVESTIGATION:

Figure 10–4 illustrates another series-parallel circuit using the same resistors. Develop a procedure for solving the currents and voltages throughout the circuit. Summarize your procedure in a laboratory report. Confirm your method by computing and measuring the voltages in the circuit.

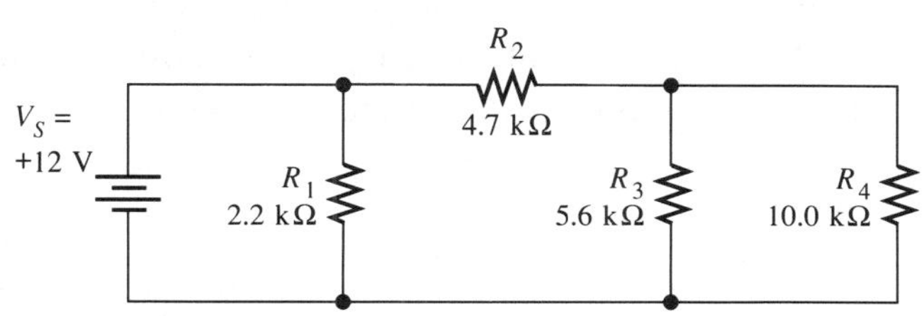

Figure 10–4

APPLICATION PROBLEM:

For many years, resistive networks have been designed to control the signal level of audio or radio frequency circuits and to match the resistance of the source and load. Circuits that perform these functions are called *attenuators,* or *pads*. There are a number of variations in pad design, but in this problem we will design an L-pad used in matching a higher resistance to a lower resistance. The circuit is shown in Figure 10–5. The first dotted box represents a signal source with a source resistance of 600 Ω. (The source resistance is internal on ac signal generators.) The L-pad consists of the two resistors shown in the second dotted box, and the load is in the third dotted box and represents the circuit being driven by the source.

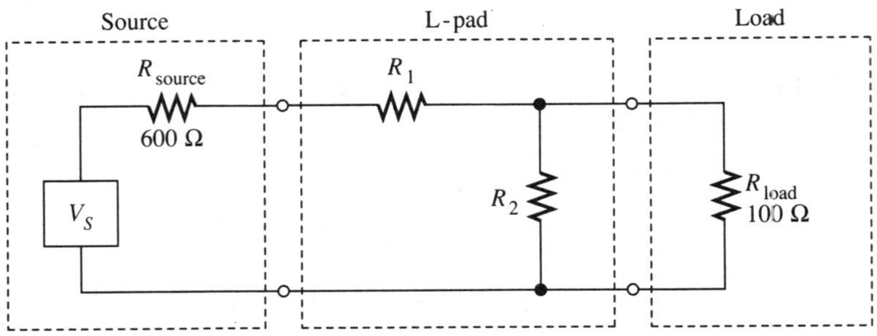

Figure 10–5

The resistors in the L-pad depend on the source resistance, the load resistance, and the desired attenuation of the pad. In this design, the attenuation must be greater than the ratio of the source to load resistance. The equations for determining these resistors are

$$R_1 = R_S\left(\frac{A-1}{A}\right) \qquad R_2 = R_S\left(\frac{1}{A - R_S/R_L}\right)$$

where: R_S = source resistance

R_L = load resistance

A = attenuation; the ratio of input/output voltage

The design requires an L-pad that matches a 600 Ω source resistance to a 100 Ω load resistance with a 10:1 attenuation from the input of the L-pad. Compute the values of the resistors in the L-pad and construct the circuit based on your design using resistors as close as possible to the calculated values. The source can be a signal generator with an internal 600 Ω resistance set for a 1 kHz sine wave or a dc power supply with a series 600 Ω resistor. Set the source voltage to 5.0 V when it is connected to the rest of the circuit. If you have correctly calculated the values of the resistors in the L-pad, the output voltage measured across the load resistor should be 0.5 V and the resistance of the circuit measured looking into the L-pad should be 600 Ω. Summarize your results in your report.

Report for Experiment 10

Name ____________________
Date ____________________
Class ____________________

OBJECTIVES:

PROCEDURE:

DATA:

Table 10–1

Component	Listed Value	Measured Value
R_1	2.2 kΩ	
R_2	4.7 kΩ	
R_3	5.6 kΩ	
R_4	10.0 kΩ	

Step 3 Equivalent Series Circuit

Table 10–2

	Computed		Measured
	Voltage Divider	Ohm's Law	
R_T			
I_T			
V_1			
$V_{2,3}$			
V_4			
I_2			
I_3			
V_T	12.0 V	12.0 V	

Step 10 Equivalent Circuit

Table 10-3

	Computed	Measured
$R_{1,2}$		
$R_{3,4}$		
R_T		
I_T		
$I_{1,2}$		
$I_{3,4}$		
V_1		
V_2		
V_3		
V_4		

RESULTS AND CONCLUSION:

FURTHER INVESTIGATION RESULTS:

APPLICATION PROBLEM RESULTS:

EVALUATION AND REVIEW QUESTIONS:

1. The voltage divider rule was developed for a series circuit, yet it was applied to the circuit in Figure 10–2.
   (a) Explain.

   (b) Could the voltage divider rule be applied to the circuit in Figure 10–3? Explain your answer:

2. As a check on your solution of the circuit in Figure 10–3, apply Kirchhoff's voltage law to each of two separate paths around the circuit. Show the application of the law.

3. Show the application of Kirchhoff's current law to the junction of R_2 and R_4 of the circuit in Figure 10–3.

4. In the circuit of Figure 10–3, assume you found that I_T was the same as the current in R_3 and R_4.
   (a) What are the possible problems?

   (b) How would you isolate the specific problem?

5. The circuit in Figure 10–6 has three *equal* resistors.
   (a) If the voltmeter reads +8.0 V, find the voltage drop across R_1. V_1 = ________
   (b) What is the source voltage? V_S = ________

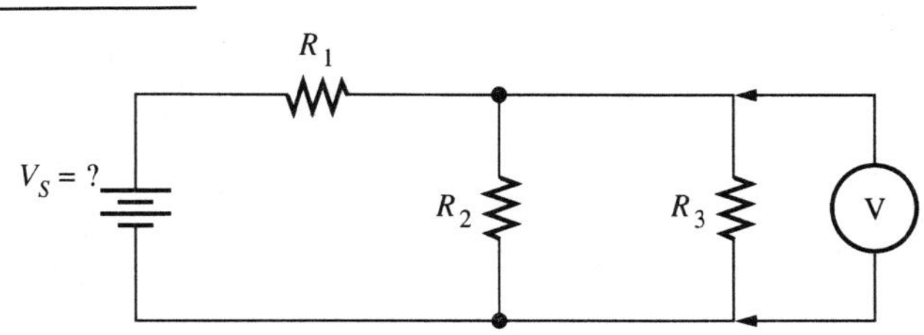

Figure 10–6

6. What basic rules determine if two resistors in a series-parallel combination circuit are connected in series or in parallel?

TECH TIP REVIEW:
Each of the following problems relate to the TECH TIP voltage-divider circuit shown in Figure 7–58 of the Floyd text. Assume a +12 V battery pack is used for the power supply as specified in the text. Determine the effect on the output for each of the following conditions.

1. Assume the +12 V battery pack is connected to the divider board between pins 1 and 3. How would this affect each of the output voltages?

2. Assume a load resistor is placed between pins 1 and 2. Which voltages rise and which voltages fall as a result of this?

3. Assume a jumper wire (short circuit) is placed between pins 3 and 4.
   (a) Which voltages rise and which voltages fall as a result of this?

   (b) Does the power supply voltage change?

11 The Superposition Theorem

PSpice

OBJECTIVES:
After performing this experiment, you will be able to:
1. Apply the superposition theorem to linear circuits with more than one voltage source.
2. Construct a circuit with two voltage sources, solve for the currents and voltages throughout the circuit, and verify your computation by measurement.

READING:
Floyd, *Principles of Electric Circuits,* Sections 8–1 through 8–4

MATERIALS NEEDED:
Resistors:
One 4.7 kΩ, one 6.8 kΩ, one 10.0 kΩ

SUMMARY OF THEORY:
To superimpose something means to lay one thing on top of another. The superposition theorem is a means by which we can solve circuits that have more than one independent voltage source. Each source is taken, one at a time, as if it were the only source in the circuit. All other sources are replaced with their internal resistance. (The internal resistance of a dc power supply or battery can be considered to be zero.) The currents and voltages for the first source are computed. The results are marked on the schematic and the process is repeated for each source in the circuit. When all sources have been taken, the overall circuit can be solved. The algebraic sum of the superimposed currents and voltages is computed. Currents that are in the same direction are added; those that are in opposing directions are subtracted, with the sign of the larger applied to the result. Voltages are treated in a like manner.

The superposition theorem will work for any number of sources as long as you are consistent in accounting for the direction of currents and the polarity of voltages. One way to keep the accounting straightforward is to assign a polarity, right or wrong, to each component. Tabulate any current that is in the same direction as the assignment as a positive current and any current that opposes the assigned direction as a negative current. When the final algebraic sum is completed, positive currents are in the assigned direction; negative currents are in the opposite direction of the assignment. In the process of replacing a voltage source with its zero internal resistance, you may completely short out a resistor in the circuit. If this occurs, there will be no current in that resistor for this part of the calculation. The final sum will still have the correct current.

PROCEDURE:
1. Obtain the resistors listed in Table 11–1. Measure each resistor and record the measured value in Table 11–1.

2. Construct the circuit shown in Figure 11–1. This circuit has two voltage sources connected to a common reference ground.

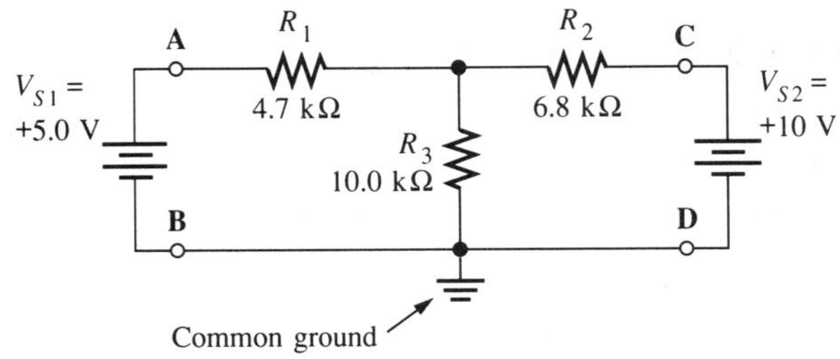

Figure 11–1

3. Remove the 10 V source and place a jumper between the points labeled **C** and **D,** as shown in Figure 11–2. This jumper represents the internal resistance of the 10 V power supply.

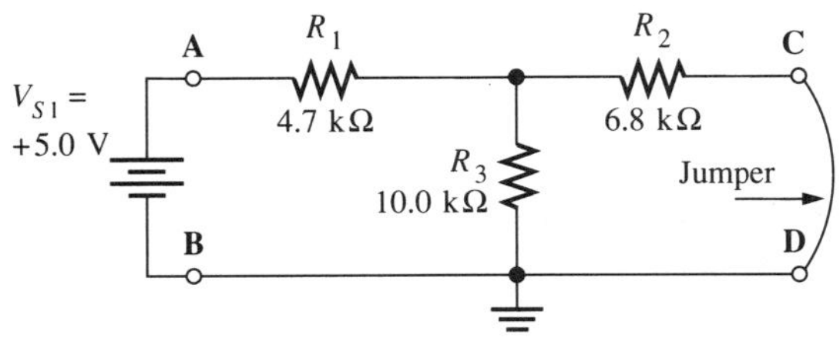

Figure 11–2

4. Compute the total resistance, R_T, seen by the +5.0 V source. Then remove the +5.0 V source and measure the resistance between points **A** and **B** to confirm your calculation. Record the computed and measured values in Table 11–2.

5. Use the source voltage, V_S, and the total resistance to compute the total current, I_T, from the +5.0 V source. This current flows through R_1, so record it as I_1 in Table 11–2. Use the current divider rule to determine the currents in R_2 and R_3. The current divider rule for I_2 and I_3 is

$$I_2 = I_T\left(\frac{R_3}{R_2 + R_3}\right) \qquad I_3 = I_T\left(\frac{R_2}{R_2 + R_3}\right)$$

Record all three currents as *positive* values in Table 11–2. This will be the assigned direction of current flow. Mark the magnitude and direction of the current on Figure 11–2.

6. Use the currents computed in step 5 and the measured resistances to calculate the expected voltage across each resistor of Figure 11–2. Then connect the +5.0 V power supply and measure the actual voltages present in this circuit. Record the computed and measured voltages in Table 11–2. Since all currents in step 5 were considered *positive*, all voltages in this step are also *positive*.

7. Remove the +5.0 V source from the circuit and move the jumper from between points **C** and **D** to between points **A** and **B.** Compute the total resistance between points **C** and **D.** Measure the resistance to confirm your calculation. Record the computed and measured resistance in Table 11–2.

8. Compute the current through each resistor in Figure 11–3. Note that the total current flows through R_2 and divides between R_1 and R_3. Mark the magnitude and direction of the current on Figure 11–3. *Important:* Record the current as a *positive* current if it is in the same direction as recorded in step 5 and as a *negative* current if it is in the opposite direction as in step 4.

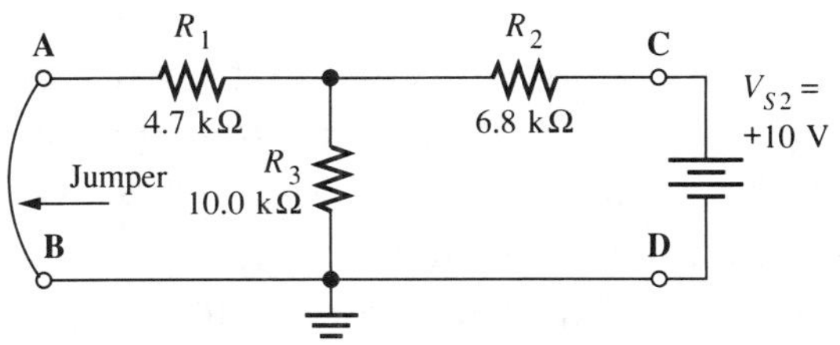

Figure 11–3

9. Use the currents computed in step 8 and the measured resistances to compute the voltage drops across each resistor. If the current through a resistor was a *positive* current, record the resistor's voltage as a *positive* voltage. If a current was a *negative* current, record the voltage as a *negative* voltage. Then connect the +10 V source, as illustrated in Figure 11–3, and measure the voltages. The measured voltages should confirm your calculation.

10. Compute the algebraic sum of the currents and voltages listed in Table 11–2. Enter the computed sums in Table 11–2. Then replace the jumper between **A** and **B** with the +5.0 V source, as shown in the original circuit in Figure 11–1. Measure the voltage across each resistor in this circuit. The measured voltages should agree with the algebraic sums. Record the measured results in Table 11–2.

FOR FURTHER INVESTIGATION:
Compute the power dissipated in each resistor in Figures 11–1, 11–2, and 11–3. Using the computed results, find out if the superposition theorem is valid for power. Summarize your computations and write a conclusion of this investigation.

APPLICATION PROBLEM:
Sometimes it is necessary to know if a voltage is larger than some reference voltage. A comparator is a circuit that can provide an indication of the relative difference between two voltages. Sometimes it is necessary to obtain the reference voltage from a voltage divider using both positive and negative supplies, as shown in Figure 11–4. The comparator looks like a very high resistance load on the divider. In this case the reference voltage is variable and can be set from −5.0 V to +5 V. (Can you verify this using the superposition theorem?)

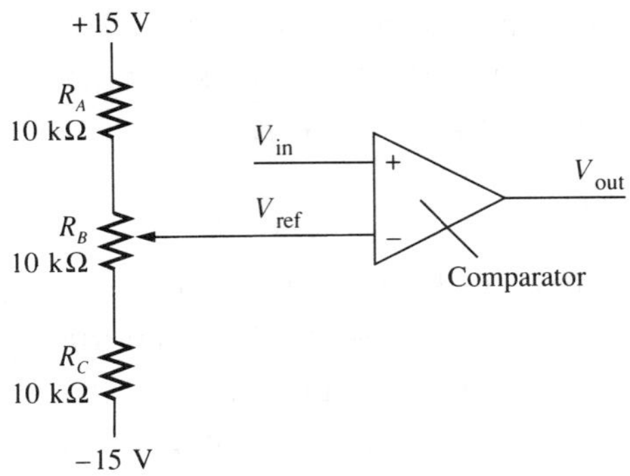

Figure 11–4

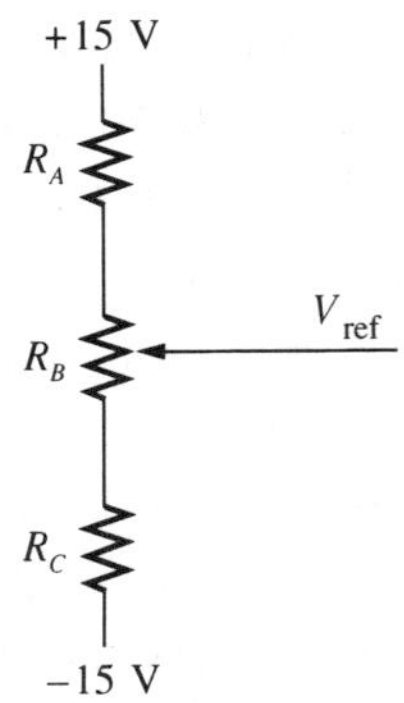

Figure 11–5

For this problem, a voltage divider is needed that operates from both +15 V and −15 V, as shown in Figure 11–5. The output reference voltage (V_{ref}) needs to be variable between −8.4 V and +5.5 V. You can use two of the resistors from this experiment for the fixed resistors and a 10 kΩ potentiometer for the variable resistor. Specify the resistors for the voltage divider shown in Figure 11–5 that meets the requirements stated. Construct the voltage divider from your design and measure the minimum and maximum voltages.

PSpice EXAMPLE:

The following source file listing is for the circuit in Figure 11−1. The output file will show the superposition current in each resistor and the node voltages for the circuit.

```
LAB11 FIG 11−1
VS1 1 0 5.0
VS2 3 0 10
R1 1 2 4.7K
R2 2 3 6.8K
R3 2 0 10K
.OP
.DC VS1 5 5 1
.PRINT DC I(R1) I(R2) I(R3)
.OPT NOPAGE
.END
```

Report for Experiment 11

Name ____________________
Date ____________________
Class ____________________

OBJECTIVES:

PROCEDURE:

DATA:

Table 11–1

Component	Listed Value	Measured Value
R_1	4.7 kΩ	
R_2	6.8 kΩ	
R_3	10.0 kΩ	

Table 11–2

	Resistance		Computed Current			Computed Voltage			Measured Voltage		
	Computed	Measured	I_1	I_2	I_3	V_1	V_2	V_3	V_1	V_2	V_3
Step 4											
Step 5											
Step 6											
Step 7											
Step 8											
Step 9											
Step 10	Totals →										

RESULTS AND CONCLUSION:

FURTHER INVESTIGATION RESULTS:

APPLICATION PROBLEM RESULTS:

EVALUATION AND REVIEW QUESTIONS:

1. (a) Prove that Kirchhoff's voltage law is valid for the circuit in Figure 11–1. Do this by substituting the measured algebraic sums from Table 11–2 into a loop equation written around the outside loop of the circuit.

   (b) Prove Kirchhoff's current law is valid for the circuit of Figure 11–1 by writing an equation showing that the currents entering a junction are equal to the currents leaving the junction. Keep the assigned direction of current from step 5 and use the signed currents computed in step 10.

2. If an algebraic sum in Table 11–2 is negative, what does this indicate?

3. What is the effect on the final result if you had been directed to record all currents in step 5 as negative currents instead of positive currents?

4. In your own words, list the steps required to apply the superposition theorem.

5. Use the superposition theorem to find the current in R_2 in Figure 11–6.

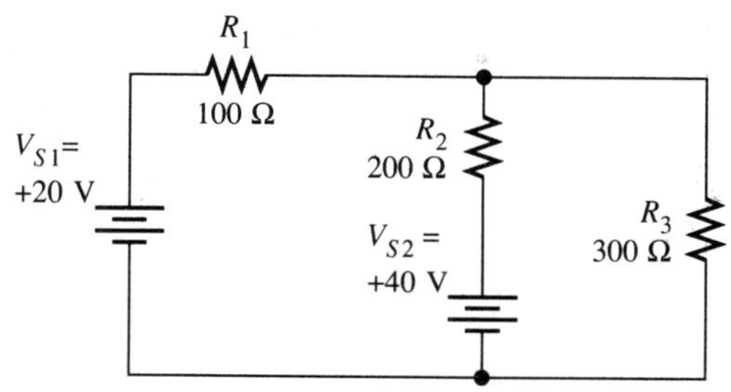

Figure 11–6

6. When the superposition theorem is applied to a circuit containing a current source, the current source is replaced by an open circuit when solving for the other sources. Why?

12 Thevenin's Theorem

OBJECTIVES:
After performing this experiment, you will be able to:
1. Change a linear resistive network into an equivalent Thevenin circuit.
2. Prove the equivalency of the network in objective 1 with the Thevenin circuit by comparing the effects of various load resistors.

READING:
Floyd, *Principles of Electric Circuits,* Section 8–5

MATERIALS NEEDED:
Resistors:
One 150 Ω, one 270 Ω, one 470 Ω, one 560 Ω, one 680 Ω, one 820 Ω
One 1 kΩ potentiometer
For Further Investigation: LED, resistors specified by student

SUMMARY OF THEORY:
In Experiment 10, you solved series-parallel circuits by developing equivalent circuits. Equivalent circuits simplify the task of solving for current and voltage in a network. The concept of equivalent circuits is basic to solving many problems in electronics.

Thevenin's theorem provides a means of reducing a complicated, linear network into an equivalent circuit when there are two terminals of special interest (usually the output). The equivalent Thevenin circuit is composed of a voltage source and a series resistor. (In ac circuits, the resistor may be represented by opposition to ac called *impedance*.) Imagine a complicated network containing multiple voltage sources, current sources, and resistors such as that shown in Figure 12–1(a). Thevenin's theorem can reduce this to the equivalent circuit shown in Figure 12–1(b). The circuit in Figure 12–1(b) is called a Thevenin circuit. A device connected to the output is a *load* for the Thevenin circuit. The two circuits have identical responses to any load!

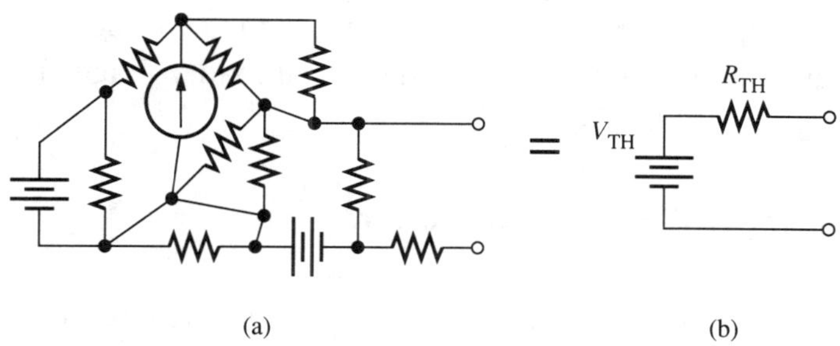

Figure 12–1

Two steps are required in order to simplify a circuit to its equivalent Thevenin circuit. The first step is to measure or compute the voltage at the terminals of interest with any load resistors removed. This open-circuit voltage is the Thevenin voltage. The second step is to compute the resistance seen at the same open terminals if sources are replaced with their internal resistance. For voltage sources, the internal resistance is usually taken as zero and for current sources the internal resistance is infinite (open circuit).

PROCEDURE:

1. Measure and record the resistance of the 6 resistors listed in Table 12–1. The last three resistors represent different load resistors that will be tested in the experiment.

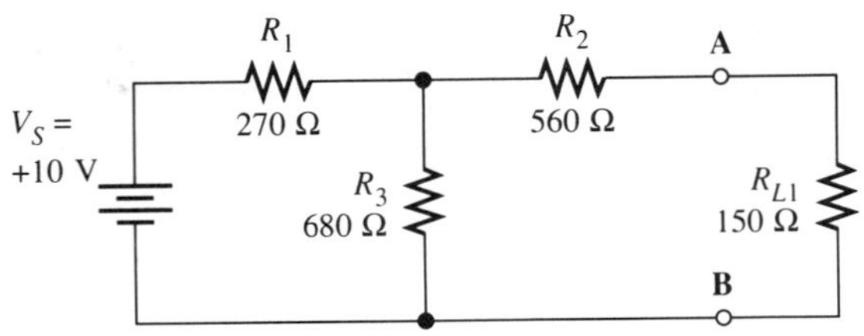

Figure 12–2

2. Construct the circuit shown in Figure 12–2. Calculate an equivalent circuit seen by the voltage source. Figure 12–3 shows equivalent circuits seen by the voltage source. Use the method illustrated in Experiment 10 and the equivalent circuits shown here to compute the expected voltage across the load resistor, V_{L1}. Do not use Thevenin's theorem at this time. Show your computation of the load voltage in the space provided in the report.

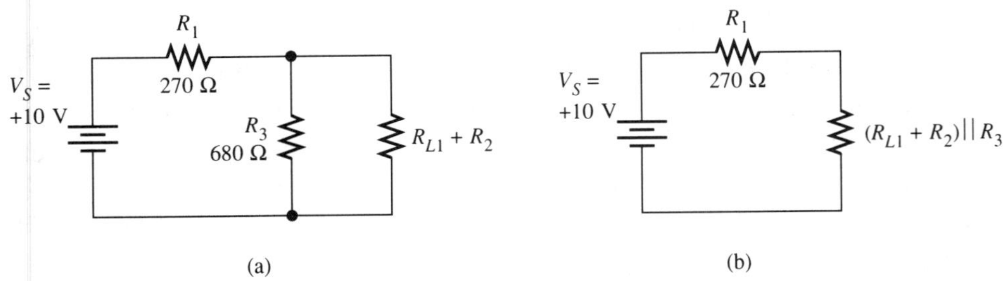

Figure 12–3

3. Measure the load voltage to verify your calculation. Enter the computed and measured load voltage in Table 12–2. They should agree within experimental uncertainty.

4. Replace R_{L1} with R_{L2}. Compute the expected voltage, V_{L2}, across the load resistor in the same manner as before. Then measure the actual load voltage. Enter the computed and measured voltage in Table 12–2.

5. Repeat step 4 using R_{L3} for the load resistor.

6. Remove the load resistor from the circuit. Calculate the open circuit voltage at the **AB** terminals. This open circuit voltage is the *Thevenin voltage* for this circuit. Record the open circuit voltage in Table 12–2 as V_{TH}.

7. Mentally replace the voltage source with a short (0 Ω). Compute the resistance between the **AB** terminals. This is the computed *Thevenin resistance* for this circuit. Then disconnect the voltage source and replace it with a jumper. Measure the actual

Thevenin resistance of the circuit. Record your computed and measured Thevenin resistance in Table 12–2.

8. In the space provided in the report, draw the Thevenin equivalent circuit. Show on your drawing the measured Thevenin voltage and resistance.

9. For the circuit you drew in step 8, compute the voltage you expect across each of the three load resistors. Since the circuit is a series circuit, the voltage divider rule will simplify the calculation. Enter the computed voltages in Table 12–3.

10. Construct the Thevenin circuit you drew in step 8. Use a 1 kΩ potentiometer to represent the Thevenin resistance. Set it for the resistance shown on your drawing. Set the voltage source for the Thevenin voltage. Place each load resistor, one at a time, on the Thevenin circuit and measure the load voltage. Enter the measured voltages in Table 12–3.

11. Remove the load resistor from the Thevenin circuit. Find the open circuit voltage with no load. Enter this voltage as the computed and measured V_{TH} in Table 12–3. Enter the measured setting of the potentiometer as R_{TH} in Table 12–3.

FOR FURTHER INVESTIGATION:

Sometimes it is useful to compute a Thevenin equivalent circuit when it is not possible to measure the Thevenin resistance directly. A simple method is to use a variable resistor as a load resistor and adjust it until the load voltage has dropped to one-half the open-circuit voltage. The variable load resistor and the internal Thevenin resistance of the source will then be equal.

The preceding method requires a variable resistor. A more general method, using a fixed resistor, is as follows:

1. Measure the no-load voltage, V_{NL}, from the generator.
2. Add a load resistor, R_L, (330 Ω is satisfactory) and measure the load voltage, V_L.
3. Calculate the load current from Ohm's law, $I_L = V_L/R_L$.
4. Compute $R_{GEN} = (V_{NL} - V_L)/I_L$.

Use one of these two methods to measure the Thevenin resistance of your function generator. Report your results, describe the method you used, and compare your measured resistance to the accepted value of generator resistance.

APPLICATION PROBLEM:

The circuit in Figure 12–4 presents an interesting problem in which the application of Thevenin's theorem can provide a ready solution. The circuit is similar to circuits used for bipolar transistor biasing although the particulars for this problem are different. The output of the voltage divider is connected to a light-emitting diode (LED), as indicated in Figure 12–4. The LED allows current to flow in only one direction and drops approximately 1.65 V when it is conducting. Given the circuit, it is a simple matter to determine the current in the diode by applying Thevenin's theorem, as shown in Figure 12–5. The current is found by subtracting 1.65 V from the Thevenin voltage (due to the LED voltage drop) and dividing by the Thevenin resistance of the circuit.

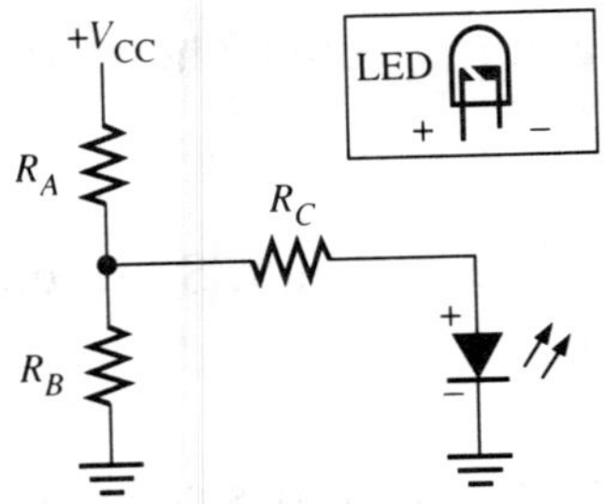

Figure 12–4

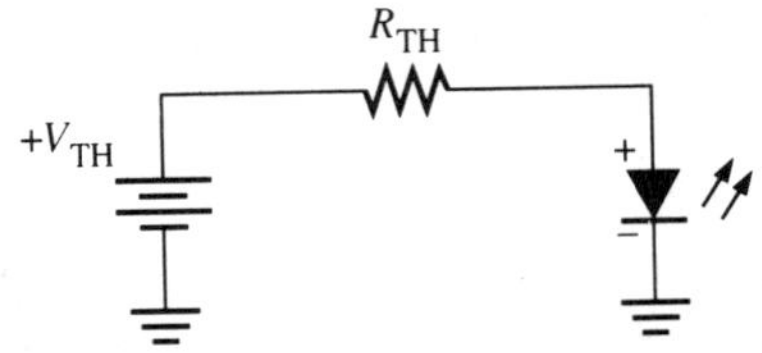

$$V_{TH} = V_{CC}\left(\frac{R_B}{R_A + R_B}\right)$$

$$R_{TH} = R_A \| R_B + R_C$$

Figure 12–5

The question is, "Can you reverse the procedure?" That is, given a required Thevenin resistance, can you find an equivalent circuit using the divider and series resistor? The problem is illustrated in Figure 12–6. The required Thevenin resistance is 600 Ω and the required current in the LED is 12 mA. R_C is given as 270 Ω and the supply voltage is +15 V. Calculate values for R_A and R_B that meet these conditions. Then construct the circuit, measure the voltages across each resistor, and prove that your design meets these requirements. Incidentally, three of the fixed resistors used in this experiment can be used to meet the design requirements. Summarize your findings in your lab report.

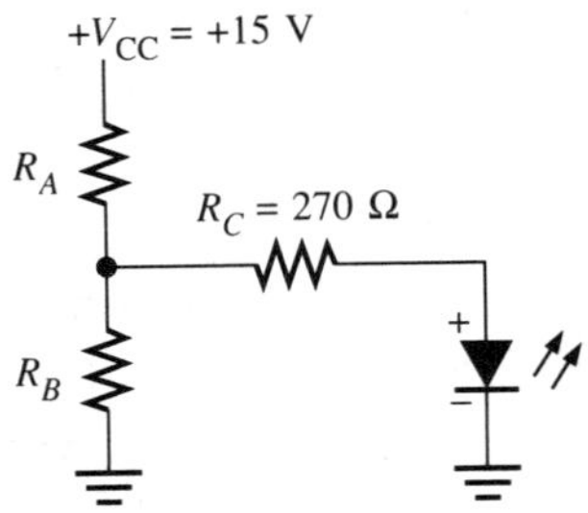

Figure 12–6

Report for Experiment 12

Name ______________
Date ______________
Class ______________

OBJECTIVES:

PROCEDURE:

DATA:

Table 12–1

Component	Listed Value	Measured Value
R_1	270 Ω	
R_2	560 Ω	
R_3	680 Ω	
R_{L1}	150 Ω	
R_{L2}	470 Ω	
R_{L3}	820 Ω	

Load Voltage Calculation (Step 2)

Table 12–2

	Computed	Measured
V_{L1}		
V_{L2}		
V_{L3}		
V_{TH}		
R_{TH}		

Table 12–3

	Computed	Measured
V_{L1}		
V_{L2}		
V_{L3}		
V_{TH}		
R_{TH}		

Thevenin Circuit (Step 8)

RESULTS AND CONCLUSION:

FURTHER INVESTIGATION RESULTS:

APPLICATION PROBLEM RESULTS:

EVALUATION AND REVIEW QUESTIONS:

1. Compare the measured voltages in Tables 12–2 and 12–3. What conclusion can you draw about the two circuits?

2. Compute the load current you would expect to measure if the load resistor in Figure 12–2 were replaced with a short circuit. (Note that this current is the Norton current, discussed in Experiment 14.) Then repeat the computation for the Thevenin circuit you drew in step 8.

3. What advantage does Thevenin's theorem offer for computing the load voltage across each of the load resistors tested in this experiment?

4. Figure 12–7(a) shows a circuit and 12–7(b) shows its equivalent Thevenin circuit. Explain why the R_1 has no effect on the Thevenin circuit.

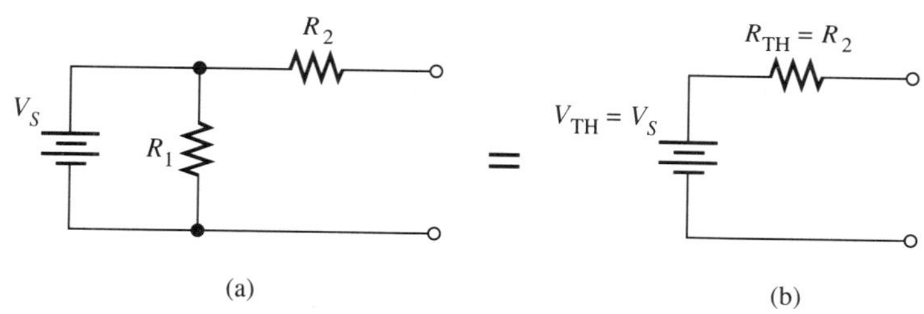

Figure 12–7

5. Refer to your results in Experiment 5. Why do you think the peak power occurred where it did?

6. Draw the Thevenin circuit for the circuits shown in Figure 12–8(a) and (b).

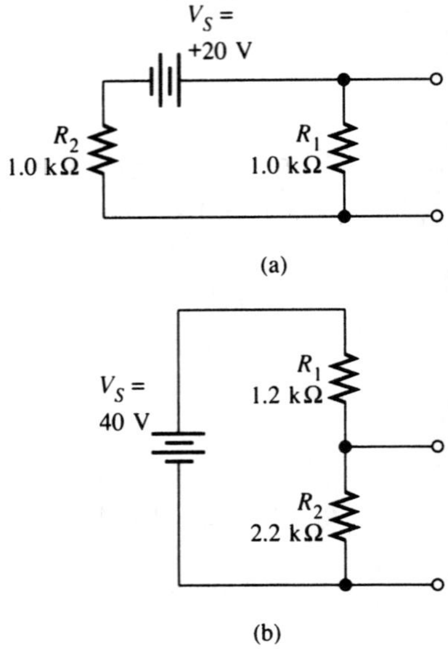

Figure 12–8

13 The Wheatstone Bridge

PSpice

OBJECTIVES:

After performing this experiment, you will be able to:

1. Calculate the equivalent Thevenin circuit for a Wheatstone bridge circuit.
2. Verify experimentally that the Thevenin circuit determined in objective 1 produces the same response to a load as the original circuit.
3. Balance a Wheatstone bridge and show the Thevenin circuit for the balanced bridge.

READING:

Floyd, *Principles of Electric Circuits,* Sections 8–5, 8–10, and 8–11

MATERIALS NEEDED:

Resistors:

One 470 Ω, one 1 kΩ, one 1.2 kΩ, one 2.2 kΩ

One 10 kΩ potentiometer

For Further Investigation: Wheatstone bridge sensitive to 0.1 Ω (optional)

Application problem: One ammeter (0–10 mA)

One decade resistance box

SUMMARY OF THEORY:

The Wheatstone bridge is a circuit used primarily in measurement applications because it can accurately compare the resistance of an unknown resistor with that of known standard resistors. The unknown resistance is frequently a transducer such as a strain gauge in which very small changes in resistance can be related to stress. The Wheatstone bridge circuit is shown in Figure 13–1(a).

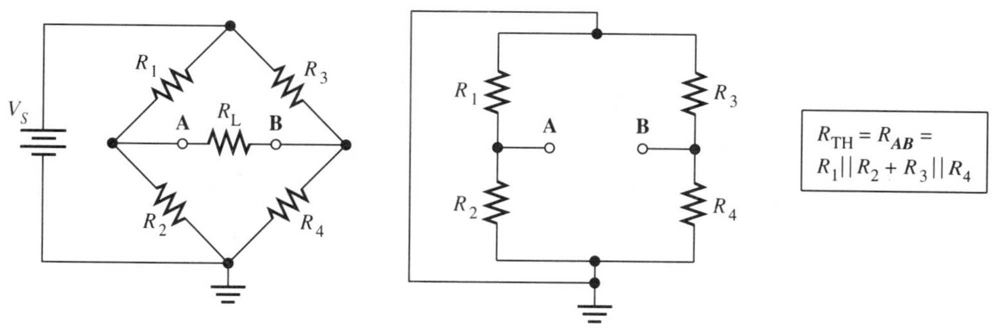

(a) (b) (c)

Figure 13–1

Thevenin's theorem can be used to analyze the current in the load resistor of the bridge. The method of applying Thevenin's theorem to the bridge is described in Floyd's text and reviewed in abbreviated form here. First, we will find the Thevenin resistance by removing the load and replacing the voltage source with a short, as shown in Figure 13–1(b). Because of this short, we see that R_1 is in parallel with R_2 and R_3 is in parallel with R_4. The Thevenin resistance is given in Figure 13–1(c).

The Thevenin voltage is the voltage that appears on the output terminal with no load. With the load removed, the bridge reduces to two opposing voltage dividers, as shown in Figure 13–2(a). The output voltage is $V_A - V_B$ and is found as in Figure 13–2(b).

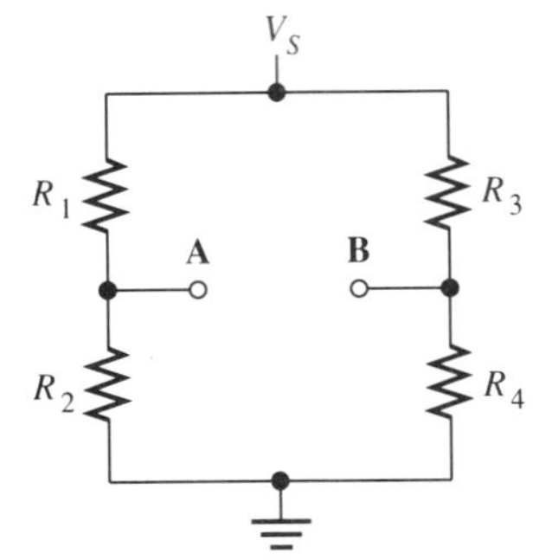

(a)

$$V_{TH} = V_A - V_B = V_S\left(\frac{R_2}{R_1 + R_2}\right) - V_S\left(\frac{R_4}{R_3 + R_4}\right)$$

(b)

Figure 13–2

PROCEDURE:

1. Measure and record the resistance of each of the four resistors listed in Table 13–1. R_4 is a 10 kΩ potentiometer. Set it for its maximum resistance and record this value.

Unbalanced Wheatstone Bridge, No Load:

2. Construct the Wheatstone bridge circuit shown in Figure 13–3. Compute the Thevenin resistance for the bridge using the method outlined in Figure 13–1. Enter the computed Thevenin resistance in Table 13–2. Then replace V_S with a short and measure R_{TH}. Record the measured Thevenin resistance in Table 13–2 and in the space provided in Figure 13–6.

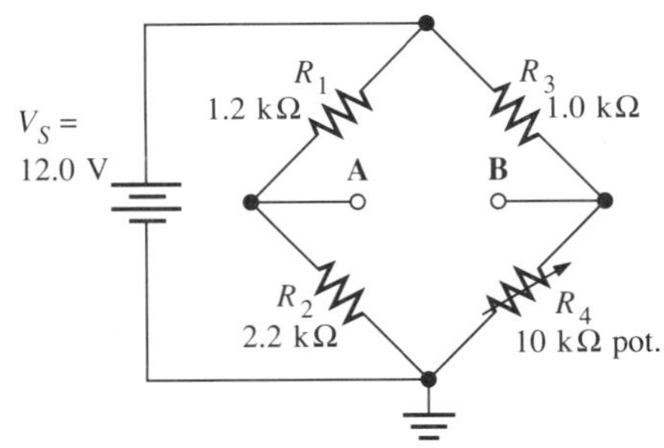

Figure 13–3

3. Compute the Thevenin voltage for the bridge using the method outlined in the Summary of Theory. Enter the computed Thevenin voltage in Table 13–2. Then measure the Thevenin voltage with your DMM. The Thevenin voltage is measured between the **A-B** terminals. Enter the measured Thevenin voltage in Table 13–2 and in the space provided in Figure 13–6 of the Report section. (The negative side of the voltage source in Figure 13–6 has been drawn on top because V_B is greater than V_A.)

Unbalanced Wheatstone Bridge, with Load:

4. Draw in the load resistor, R_L, between the **A-B** terminals in the Thevenin circuit of Figure 13–6. Show the measured value of R_L. Compute the expected voltage drop, V_L, across the load resistor. Enter the computed voltage drop in Table 13–2.

5. Place the load resistor across the **A-B** terminals of the bridge circuit (Figure 13–3) and measure the load voltage, V_L. If the measured value does not agree with the computed value, recheck your work.

Balanced Wheatstone Bridge:

6. Monitor the voltage across the load resistor and carefully adjust R_4 until the bridge is balanced. (How can you tell?) When balance is achieved, record the load voltage reading on the last line in Table 13–3. Then remove the load resistor and measure the output voltage between **A** and **B.** Since the load resistor has been removed, this measurement represents the Thevenin voltage of the balanced bridge. Enter the measured Thevenin voltage in Table 13–3 and in Figure 13–7.

7. Replace the voltage source with a short. With the short in place, measure the Thevenin resistance, as before. Enter the measured resistance in Table 13–3 and in Figure 13–7.

FOR FURTHER INVESTIGATION:

To do this investigation, you will need a calibrated Wheatstone bridge, capable of making resistance measurements within 0.1 Ω or better. A Wheatstone bridge can be used to determine the location of a short to ground in a multiple-conductor cable. The Wheatstone bridge is connected to make a ratio measurement. Simulate a multiple-conductor cable with two small-diameter wires (#24 gauge or higher) of at least 150 ft in length. You will need an accurate total resistance of the wire, which you can obtain from either the Wheatstone bridge or a sensitive ohmmeter. Place a short to ground on one of the wires at some arbitrary location along the wire. See Figure 13–4 for a diagram. The wire will form two of the legs of a Wheatstone bridge, as illustrated.

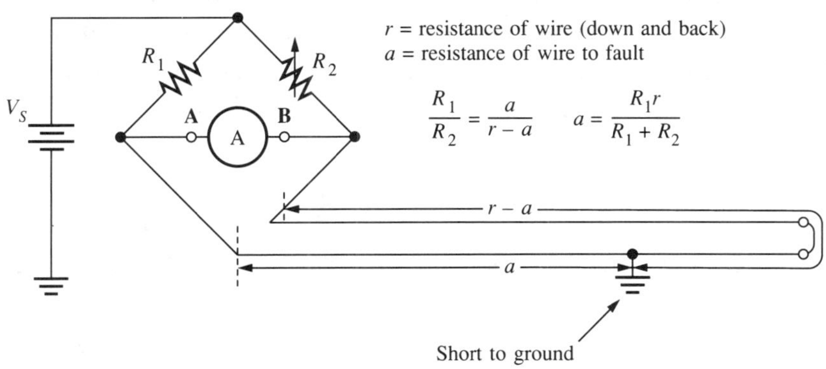

Figure 13–4

Call the total resistance of the wire r (down and back) and the resistance of the wire to the fault a. The bridge is balanced, and the resistance a is determined by the equation shown in Figure 13–4. The fractional distance to the fault is the ratio of a to ½r. If you know the total length of the wire, you can find the distance to the fault by setting up a proportion. Investigate this and report on your results.

APPLICATION PROBLEM:

Wheatstone bridge circuits find applications in a number of measurement circuits. One way of converting resistance to a current or voltage reading is to use an unbalanced Wheatstone bridge with a resistive sensor in one (or more) of the arms. A thermistor is a temperature-sensitive resistor that usually has a very nonlinear graph of temperature versus resistance characteristic. By using the loading effects of an unbalanced Wheatstone bridge, this nonlinear characteristic can be converted to a fairly linear thermometer over a selected temperature range.

In this application problem, you need to calibrate an ammeter as a temperature gauge. The circuit is shown in Figure 13–5. The circuit is designed to read temperatures from 40°C to 100°C. R_4 represents a thermistor with the resistance-temperature characteristic given in Table 13–4. Use a decade resistance box to represent the thermistor. (A potentiometer may be used if a decade resistance box is unavailable.) Set the decade box to each resistance listed in Table 13–4 and record the observed current in Table 13–4. Then graph the temperature versus the meter current on Plot 13–1 for the simulated thermistor. For comparison, draw the resistance-temperature characteristic on the same graph. (You will need to assign two scales to the *y*-axis.)

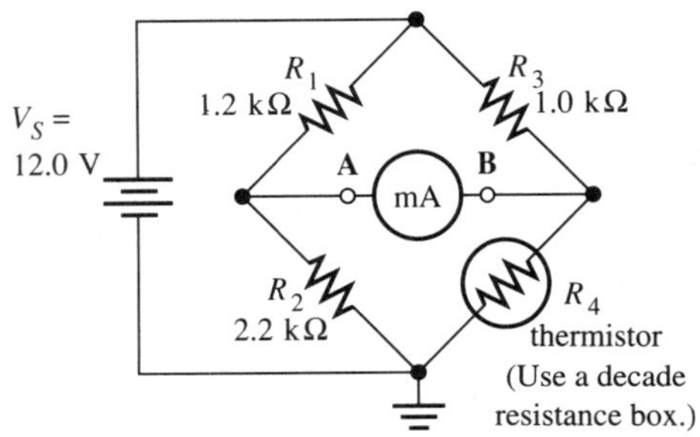

Figure 13–5

PSpice EXAMPLE:

The following PSpice examples are for the circuit in Figure 13–8 (Evaluation and Review Question 1). The first two runs show how the Thevenin voltage and resistance can be found by assigning the load resistor an extremely large value (100×10^{12} Ω). The third run uses the correct value for R_L and will give the correct node voltages for the circuit. Note that point **A** is node 2 in the listing and point **B** is node 3 in the listing.

Run 1:

```
BRIDGE CIRCUIT LAB13 FIG 13–8
V 1 0 DC 12V
R1 1 2 330
R2 2 0 330
R3 1 3 330
R4 3 0 220
RL 2 3 100E12
.TF V(2) V
.OP
.OPTIONS NOPAGE
.PRINT DC I(RL) V(RL) V(2) V(3)
.END
```

Run 2:

```
BRIDGE CIRCUIT LAB13 FIG 13–8
V 1 0 DC 12V
R1 1 2 330
R2 2 0 330
R3 1 3 330
R4 3 0 220
RL 2 3 100E12
.TF V(3) V
.OP
.OPTIONS NOPAGE
.PRINT DC I(RL) V(RL) V(2) V(3)
.END
```

Run 3:

```
BRIDGE CIRCUIT LAB13 FIG 13–8
V 1 0 DC 12V
R1 1 2 330
R2 2 0 330
R3 1 3 330
R4 3 0 220
RL 2 3 100
.DC V 12V 12V 12V
.OP
.OPTIONS NOPAGE
.PRINT DC I(RL) V(RL) V(2) V(3)
.END
```

Report for Experiment 13

Name ________________
Date ________________
Class ________________

OBJECTIVES:

PROCEDURE:

DATA:

Table 13–1

Component	Listed Value	Measured Value
R_1	1.2 kΩ	
R_2	2.2 kΩ	
R_3	1.0 kΩ	
R_L	470 Ω	
R_4	10 kΩ pot.	

The Unbalanced Wheatstone Bridge:

Table 13–2

	Computed	Measured
V_{TH}		
R_{TH}		
V_L		

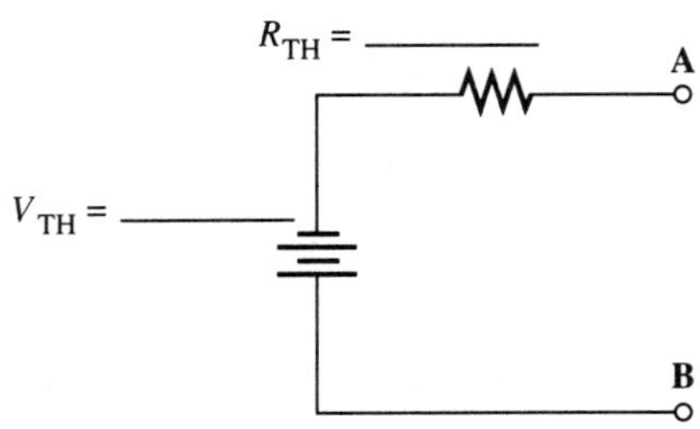

Figure 13–6

The Balanced Wheatstone Bridge:

Table 13–3

	Measured
V_{TH}	
R_{TH}	
V_L	

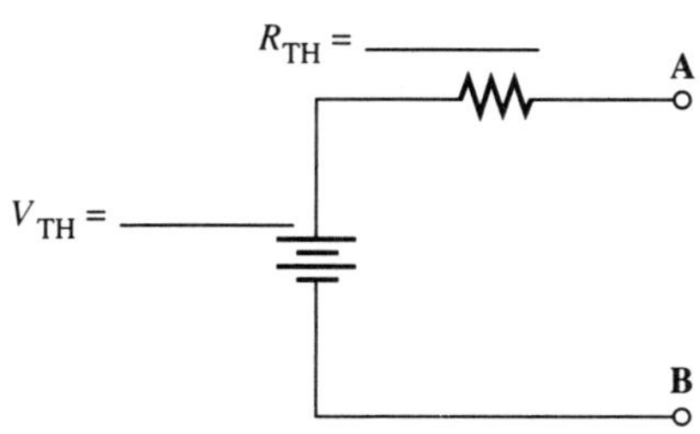

Figure 13–7

RESULTS AND CONCLUSION:

FURTHER INVESTIGATION RESULTS:

APPLICATION PROBLEM RESULTS:

Table 13–4

Temperature (°C)	Thermistor Resistance (kΩ)	Meter Current (mA)
40°	1.943 kΩ	
50°	1.374 kΩ	
60°	0.993 kΩ	
70°	0.731 kΩ	
80°	0.547 kΩ	
90°	0.416 kΩ	
100°	0.322 kΩ	

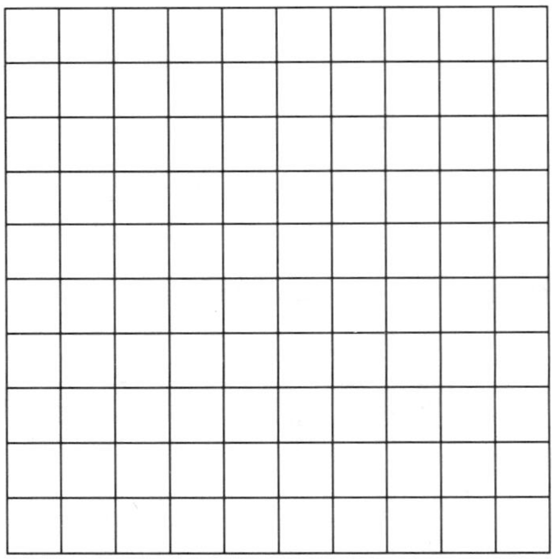

Plot 13–1

EVALUATION AND REVIEW QUESTIONS:

1. Compute the load current for the bridge shown in Figure 13–8. Show your work.

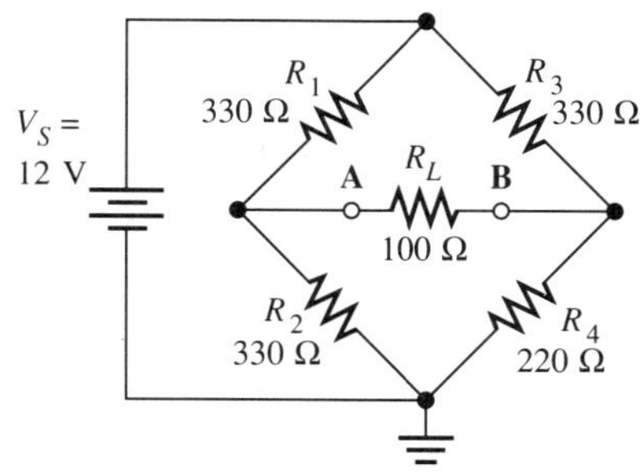

Figure 13–8

2. If you doubled the load resistor in the circuit of Figure 13–8, the load current would *not* be half as much. Why not?

3. In step 6, how did you determine that the Wheatstone bridge was balanced?

4. (a) Does a change in the load resistor change the currents in the arms of an *unbalanced* bridge? Explain your answer.

   (b) Does a change in the load resistor change the currents in the arms of a *balanced* bridge? Explain.

5. (a) What would happen to the load current of an *unbalanced* bridge if all the bridge resistors were doubled in size?

   (b) What would happen to the load current of a *balanced* bridge if all the bridge resistors were doubled in size?

6. (a) What would happen to the load current of an *unbalanced* bridge if the source voltage were doubled?

(b) What would happen to the load current of a *balanced* bridge if the source voltage were doubled?

TECH TIP REVIEW:

Each of the following questions relate to the TECH TIP circuit shown in Figure 8–67 of the Floyd text. Assume that R_2 of the Wheatstone bridge is adjusted for *balance* (output = 0 V) at a temperature of 170°F. The output voltage is measured between A and B with B as the reference. Describe what will happen to the output of the bridge for each of the following changes (output becomes positive, negative, or does not change). Changes occur independently of each other.

1. The temperature in the tank increases.

2. The power supply voltage drops to 11 V.

3. The input of the amplifier is disconnected from the bridge.

14 Norton's Theorem

OBJECTIVES:

After performing this experiment, you will be able to:

1. Set up a current source, measure the output current for several load resistors, and graph the results.
2. Given a resistive circuit, compute the equivalent Norton circuit. Construct the Norton circuit and prove that it has the same response as the original circuit.

READING:

Floyd, *Principles of Electric Circuits,* Sections 8–2 through 8–6

MATERIALS NEEDED:

Resistors:

One of each: 47 Ω, 470 Ω, 560 Ω, 1.0 kΩ, 1.2 kΩ, 1.5 kΩ, 1.8 kΩ, 2.2 kΩ

One 1 kΩ potentiometer and one 5 kΩ potentiometer

One MPF102 *n*-channel field-effect transistor

SUMMARY OF THEORY:

The most common and basic type of power supply maintains a constant output voltage over certain design limits regardless of changes in load or temperature. This is generally what is referred to when using the term *power supply.* Constant current supplies are less common. They maintain a constant current in spite of changes in load or temperature. Actual voltage sources are "loaded down" as the resistance on the output terminals decreases. Current sources are just the opposite. Current sources can easily provide current to a short but cannot supply current to an open.

One component that approximates a constant current source is the field-effect transistor (FET). A field-effect transistor is a voltage-controlled transistor that uses an electrostatic field to control current flow. The FET begins with a doped piece of silicon called a *channel.* On one end of the channel is a terminal called the *source* and on the other end of the channel is a terminal called the *drain.* Current flow in the channel is controlled by a voltage applied to a third terminal called the *gate.* The FET uses a gate *voltage* to control the current. You will learn more about field-effect transistors in another course. For this experiment, the FET should simply be thought of as a current source.

Recall that Thevenin's theorem allows us to replace any linear, two terminal network with a single voltage source and a single series resistor (or *impedance,* in the case of ac circuits). This is a very powerful theorem for analyzing the behavior of a two-terminal driving source on some load. It is also used in the analysis of a larger circuit by reducing a portion of the circuit driving two terminals of special interest. Norton's theorem is similar to Thevenin's theorem and allows us to reduce a complex circuit by replacing the driving source with a current source in parallel with a resistance. The Norton current is equal to the current that flows through a short placed on the terminals of interest and the Norton resistance is the same as the Thevenin resistance. Figure 14–1 compares the Thevenin and Norton circuits.

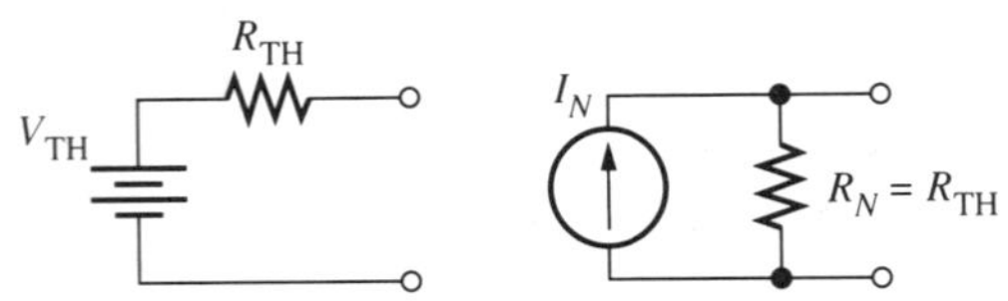

(a) Thevenin circuit (b) Norton circuit

Figure 14–1

PROCEDURE:

1. Set up the current source circuit shown in the schematic of Figure 14-2(a). Adjust R_S for maximum resistance. Set the power supply for − 15 V. The FET pin configuration and a protoboard wiring diagram are shown in Figure 14-2(b). The test resistors are added in step 3.

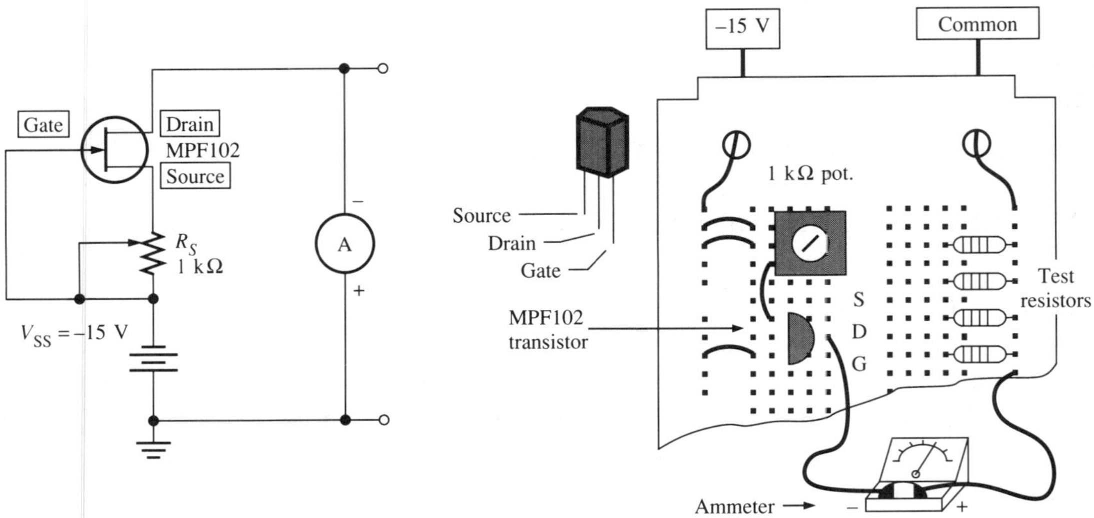

(a) Current souce schematic (b) Protoboard wiring

Figure 14–2

2. Connect a 0–10 mA ammeter between the current source leads as shown. Adjust R_S until the meter reads 5.0 mA. We will ignore the internal resistance of the meter. The 5.0 mA is entered in Table 14–1 for $R_S = 0$.

3. Measure each test resistor listed in Table 14–1 and record its measured resistance in the table. Then, one at a time, connect each test resistor in series with the ammeter by moving the wire connected to the + terminal of the ammeter to each resistor. Do not adjust the potentiometer. Read and record the current through each resistor in Table 14–1. You should observe that your current source is approximately constant for all of the test resistors.

4. On Plot 14–1, graph the resistance as a function of current for the data you took in step 3 for the current source.

5. In this step you will add a circuit on the other end of the same protoboard. Do not remove the current source circuit that you constructed in step 1. Measure and record the resistance of each resistor listed in Table 14–2. Then construct the T circuit shown in Figure 14–3.

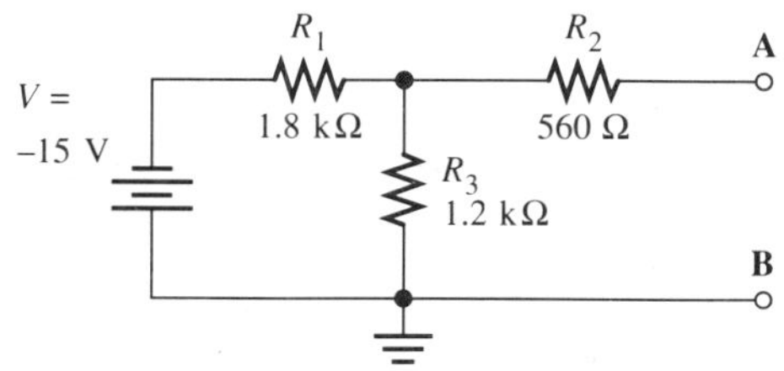

Figure 14–3

6. Compute a Norton equivalent at terminals **A** and **B** for the T circuit in Figure 14–3. Use your measured resistors to compute the Norton circuit. Draw the Norton circuit in the space provided in your report. Show the value of the Norton resistance and current source on your drawing.

7. A quick check that you have computed the Norton circuit correctly is to compute a Thevenin circuit and apply Norton's theorem to the result. Make this check in the space provided in the report.

8. Adjust R_S on the current source until the ammeter reads the value you computed for the Norton current. Obtain a 5 kΩ potentiometer and set the resistance for the Norton resistance computed in step 6. Place the Norton resistance in parallel with the current source, forming an equivalent Norton circuit. The circuit and protoboard wiring for this step is shown in Figure 14–4(a) and (b).

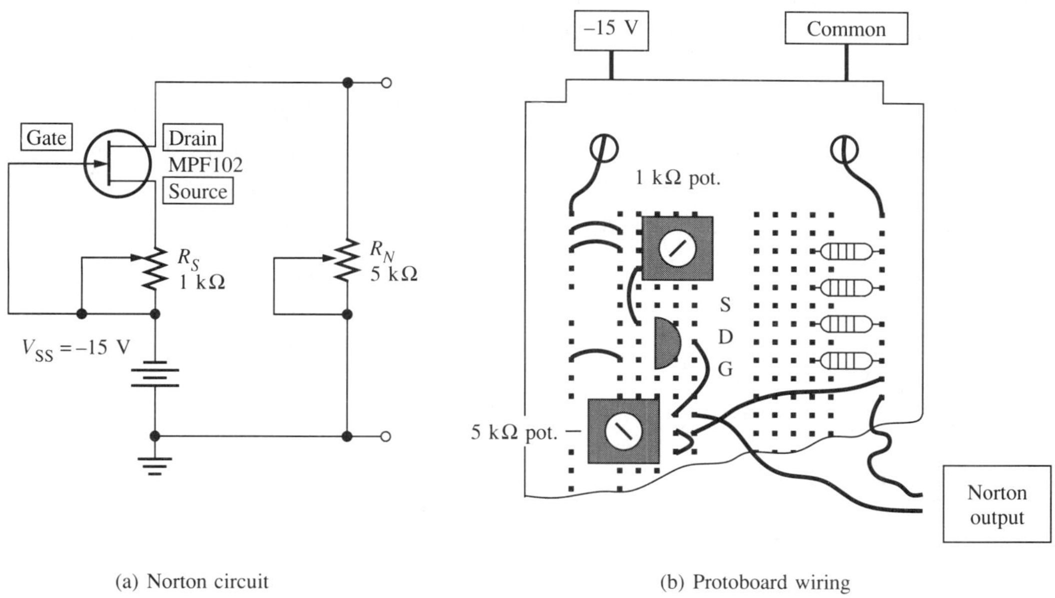

Figure 14–4

9. If you have correctly set up the Norton circuit, it will have the same output for any load as the T circuit. Remove the ammeter from the Norton circuit. Record the no-load voltage for both circuits in Table 14–3. Then place each load resistor listed in Table 14–3 across the T circuit and then across the Norton circuit. Measure the output voltage across each load resistor and record your measured voltages in Table 14–3.

FOR FURTHER INVESTIGATION:

A variation of the T circuit used in this experiment is the bridged-T circuit shown in Figure 14–5. Bridged-T networks are useful in attenuator, filter and matching circuits. Construct the circuit. Replace the power supply connections with a short and determine the Norton resistance. Find the open-circuit voltage with no load and determine the Thevenin voltage. Transform the equivalent Thevenin circuit into a Norton circuit. Adjust the Norton circuit (Figure 14–3) for the Norton current and resistance you measured. Then using the load resistors from this experiment, prove that the two circuits are equivalent.

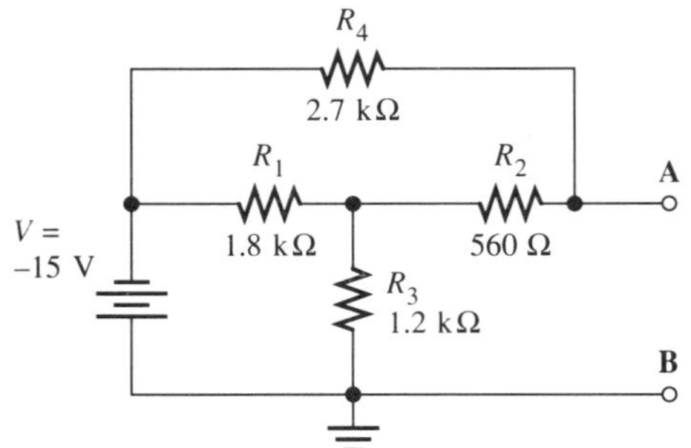

Figure 14–5

APPLICATION PROBLEM:

Norton's theorem can be used to aid in the design of voltage dividers for different loads. For example, assume you need a −5 V output from the − 15 V source voltage used in this experiment. The smallest load resistor used in this experiment was 470 Ω. The Norton circuit with this load resistor is shown in Figure 14–6(a). The current in the 470 Ω load is found from Ohm's law to be 10.6 mA. Allowing 10X more current in the Norton resistance will prevent the divider from being loaded excessively. This implies 47 Ω for the Norton resistance as shown. Notice that you can now develop two equations with two unknowns:

1. Equation 1 specifies the relation between R_N and R_A and R_B:

$$\frac{1}{R_N} = \frac{1}{R_A} + \frac{1}{R_B}$$

2. Equation 2 specifies the ratio of R_A and R_B:

$$\frac{R_B}{R_A + R_B} = \frac{5\text{ V}}{15\text{ V}}$$

Using the preceding equations, complete the design of the divider in Figure 14–6(b) by specifying the required resistors in the divider string. Specify their value and their wattage rating. Then test the load resistors used in this experiment on your design.

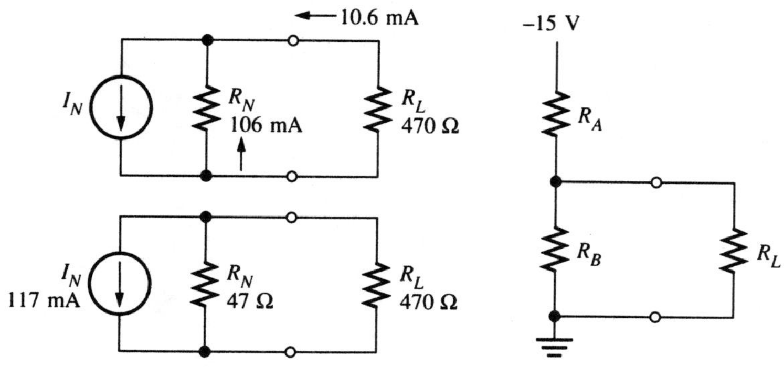

(a) Norton circuits (b) Voltage divider design

Figure 14–6

Report for Experiment 14

Name ______________________
Date ______________________
Class ______________________

OBJECTIVES:

PROCEDURE:

DATA:

Table 14–1

Resistor	Listed Value	Measured Resistance	Measured Current
$R_S = 0$			5.0 mA
R_{L1}	470 Ω		
R_{L2}	1.0 kΩ		
R_{L3}	1.5 kΩ		
R_{L4}	2.2 kΩ		

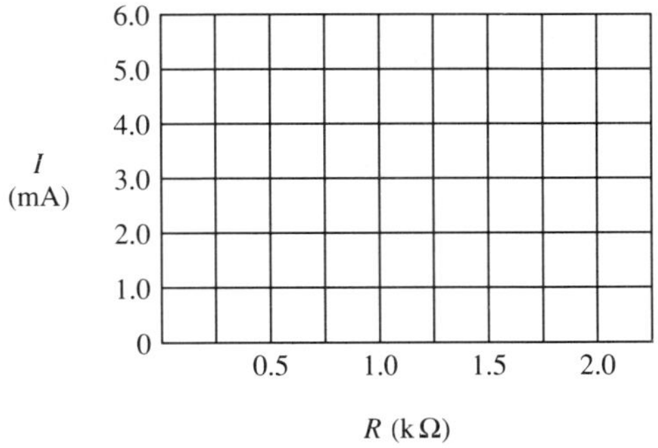

Plot 14–1

Table 14–2

Resistor	Listed Value	Measured Resistance
R_1	1.8 kΩ	
R_2	560 Ω	
R_3	1.2 Ω	

Norton Circuit

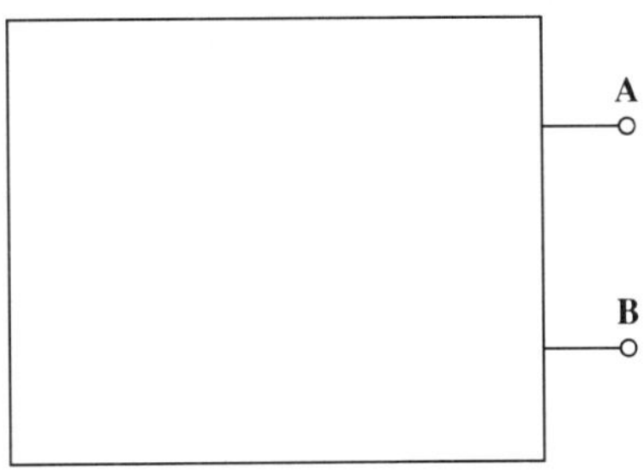

Thevenin Circuit

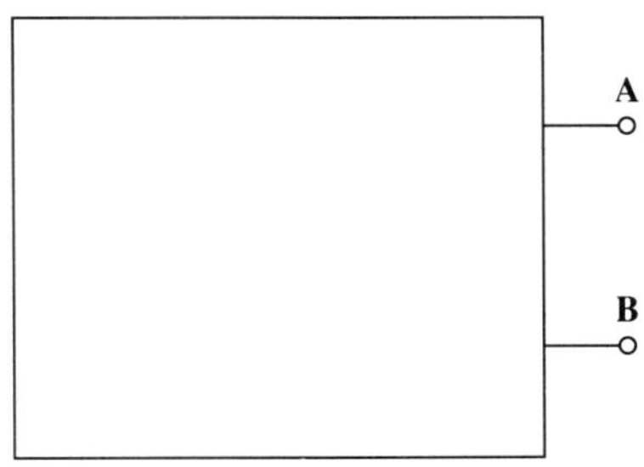

Table 14–3

Resistor	Listed Value	Measured Output Voltage	
		T Circuit	Norton Cir.
None			
R_{L1}	470 Ω		
R_{L2}	1.0 kΩ		
R_{L3}	1.5 kΩ		
R_{L4}	2.2 kΩ		

RESULTS AND CONCLUSION:

FURTHER INVESTIGATION RESULTS:

APPLICATION PROBLEM RESULTS:

EVALUATION AND REVIEW QUESTIONS:

1. (a) What is the internal resistance of an ideal current source?

   (b) What is the internal resistance of an ideal voltage source?

2. (a) How did the actual current source in this experiment differ from an ideal source?

   (b) What would you expect to happen to the output current if a 10 kΩ resistor were used as a load resistor in step 3?

3. (a) Assume a battery can deliver a maximum of 1 A into a 11 Ω load. If the open-circuit load voltage is 12 V, draw the equivalent Thevenin and Norton circuits for the battery.

4. A signal generator has a no load voltage of 10.0 V. When a 600 Ω resistor is placed across the output terminals, the voltage drops to 5.0 V. Draw the Thevenin and Norton equivalent circuits for the generator.

5. (a) A 100 V power supply has a series limiting resistor of 100 kΩ connected to its output. Draw the equivalent Norton circuit.

   (b) Explain why the circuit you drew is a good current source for resistors under 1 kΩ.

6. In Experiment 13, you found the Thevenin equivalent circuit for an unbalanced Wheatstone bridge (Figure 13–4). Draw the Norton equivalent of this circuit and use it to calculate the current in a 470 Ω load.

15 Circuit Analysis Methods

OBJECTIVES:

After performing this experiment, you will be able to:

1. Write mesh and node equations for a resistive circuit.
2. Prove, through measurement, that the equations written in objective 1 are valid.

READING:

Floyd, *Principles of Electric Circuits,* Sections 9–1 through 9–5

MATERIALS NEEDED

Resistors:

One 1.0 kΩ, one 2.0 kΩ, one 3.6 kΩ, one 4.7 kΩ, one 10 kΩ

SUMMARY OF THEORY:

Reducing a circuit by substituting equivalent parallel or series elements is a technique that works in most but not all circuits. Sometimes a circuit can have a complex group of elements which are not in either a series or parallel combination such as the bridged-T circuit shown in Figure 15–1. For such a circuit, the method of mesh equations or the method of node voltages is applicable.

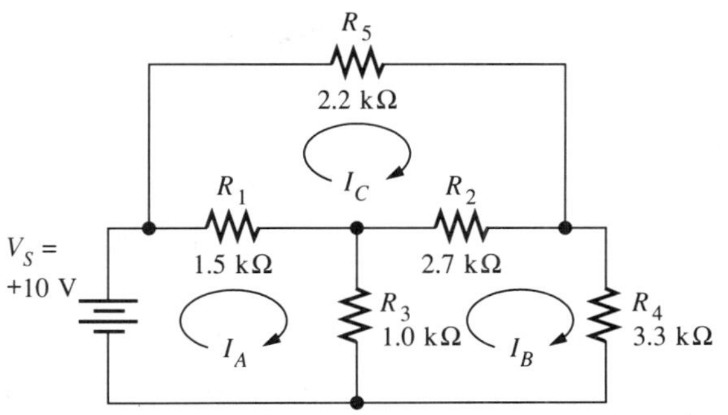

Loop **A**:
$$-V_S + (I_A - I_C)R_1 + (I_A - I_B)R_3 = 0$$

Loop **B**:
$$(I_B - I_A)R_3 + (I_B - I_C)R_2 + I_B R_4 = 0$$

Loop **C**:
$$(I_C - I_A)R_1 + I_C R_5 + (I_C - I_B)R_2 = 0$$

Figure 15–1

The mesh equation method of analysis leads to a set of simultaneous equations which can be solved by algebraic methods. The equations are written by assigning "fictitious" currents around any closed path (called loops) in a circuit. There must be at least one loop current in every branch, and a particular loop current cannot be the only loop current in two or more branches. The loop currents are indicated by an arrow that closes on itself and defines the current that flows around the perimeter of a loop. Although the direction of loop currents is arbitrary, to minimize errors it is a good idea to choose the same direction when writing loop currents (we will use the clockwise direction). Kirchhoff's voltage law is then applied to each loop forming a loop equation. The current in a branch of the circuit can be found by adding the algebraic sum of the loop currents found in that branch. For example, the equations for the loops in Figure 15–1 are shown here.

Notice that I_A is always positive in loop A, I_B is positive in loop B, and so forth. Substituting the numerical values into the preceding equations gives:

$$-10\text{ V} + (I_A - I_C)1.5\text{ k}\Omega + (I_A - I_B)1.0\text{ k}\Omega = 0$$
$$(I_B - I_A)1.0\text{ k}\Omega + (I_B - I_C)\,2.7\text{ k}\Omega + I_B\,3.3\text{ k}\Omega = 0$$
$$(I_C - I_A)1.5\text{ k}\Omega + \mathrm{I}_C\,2.2\text{ k}\Omega + (I_C - I_B)\,2.7\text{ k}\Omega = 0$$

Rearranging the equations in standard form:

$$2.5\text{ k}\Omega\, I_A - 1.0\text{ k}\Omega\, I_B - 1.5\text{ k}\Omega\, I_C = 10\text{ V}$$
$$-1.0\text{ k}\Omega\, I_A + 7.0\text{ k}\Omega\, I_B - 2.7\text{ k}\Omega\, I_C = 0\text{ V}$$
$$-1.5\text{ k}\Omega\, I_A - 2.7\text{ k}\Omega\, I_B + 6.4\text{ k}\Omega\, I_C = 0\text{ V}$$

Linear equations can be solved for using the method of determinants as shown in the text. Solving gives

$$I_A = 5.905\text{ mA}, \quad I_B = 1.645\text{ mA}, \quad \text{and} \quad I_C = 2.078\text{ mA}$$

These values can be used to find the individual branch currents:

$$I_1 = I_A - I_C = 5.905\text{ mA} - 2.078\text{ mA} = 3.827\text{ mA}$$
$$I_2 = I_B - I_C = 1.645\text{ mA} - 2.078\text{ mA} = -0.433\text{ mA}$$
$$I_3 = I_A - I_B = 5.905\text{ mA} - 1.645\text{ mA} = 4.26\text{ mA}$$
$$I_4 = I_B = 1.645\text{ mA}$$
$$I_5 = I_C = 2.078\text{ mA}$$

The negative current for I_2 indicates that it flows opposite to the direction assumed for I_B.

The node-voltage method is another method for solving circuits that cannot be reduced by equivalent circuit analysis. A node in a circuit is any point where two or more current paths come together. In the node-voltage method, the actual currents in the branches are found by setting up a set of simultaneous equations for each node. The equations are based on the Kirchhoff's current law—the sum of the currents entering a node is equal to the sum of the currents leaving a node. We'll use the same circuit as an example. It is shown in Figure 15–2 with the nodes shown as points **A, B, C,** and **D.** Let **D** = the reference node, equivalent to ground (0 V). Assume a direction of current flow for each resistor as shown in Figure 15–2. We see that point **A** is defined by the source voltage, V_S. The equations for nodes **B** and **C** are shown here.

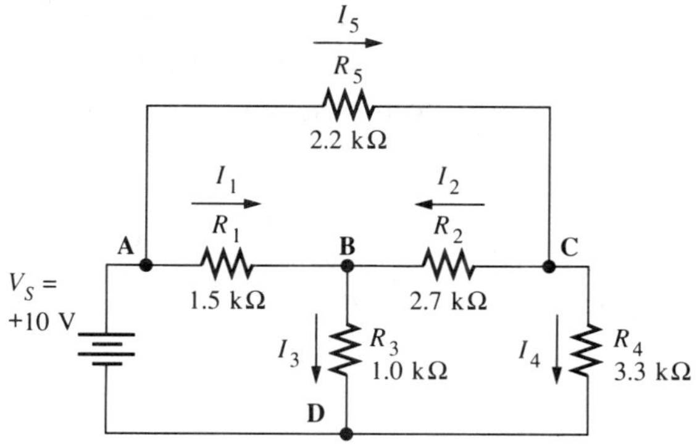

For node **B**:

$$I_1 + I_2 - I_3 = 0$$

$$\frac{V_S - V_B}{R_1} + \frac{V_C - V_B}{R_2} - \frac{V_B - V_D}{R_3} = 0$$

For node **C**:

$$-I_2 - I_4 + I_5 = 0$$

$$-\frac{V_C - V_B}{R_2} - \frac{V_C - V_D}{R_4} + \frac{V_S - V_C}{R_5} = 0$$

Figure 15–2

By substituting the known values into the equations given with Figure 15–2, the voltages at nodes **B** and **C** can be found.

PROCEDURE:

1. Measure and record the resistance of the resistors listed in Table 15–1.

2. Construct the bridged-T circuit shown in Figure 15–3. Using the mesh equation method illustrated in the Summary of Theory, write a loop equation for loops A, B, and C shown in Figure 15–3. Then, using the method of determinants illustrated in Floyd's text, solve for each of the loop currents. Show your work in the space provided in the report and enter the computed loop currents in Table 15–2.

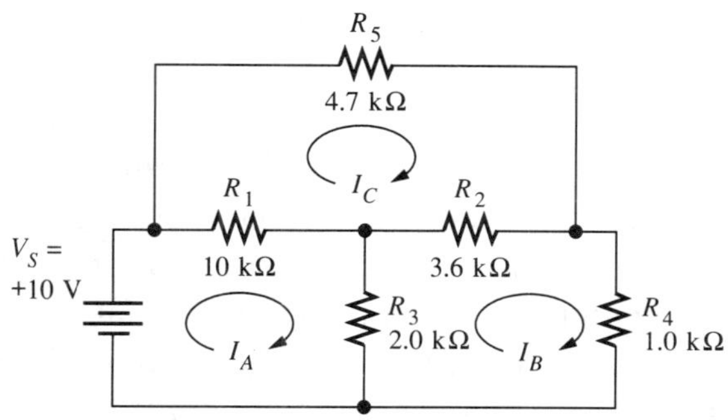

Figure 15–3

3. Using the computed loop currents, solve for the current in each resistor. Enter the computed currents in Table 15–2.

4. Using the computed currents in each resistor, apply Ohm's law to find the computed voltage drop across each resistor. Then, measure the voltage across each resistor in the circuit to confirm your calculated values are correct. Enter the computed and measured voltages in Table 15–3.

5. The circuit for this experiment has been redrawn in Figure 15–4 to identify four nodes labeled **A, B, C, D,** and the assumed direction of current into those nodes. Assume node **D** is the reference (ground) node. The voltage at node **A** is the source voltage, V_S.

Write the equations for nodes **B** and **C** in the space provided in your report. Substitute the known values into your equations and solve for the voltage at node **B** and node **C.** Your results for node **B** should be consistent with the measured voltage across R_3 and your results for node **C** should be consistent with the measured voltage across R_4.

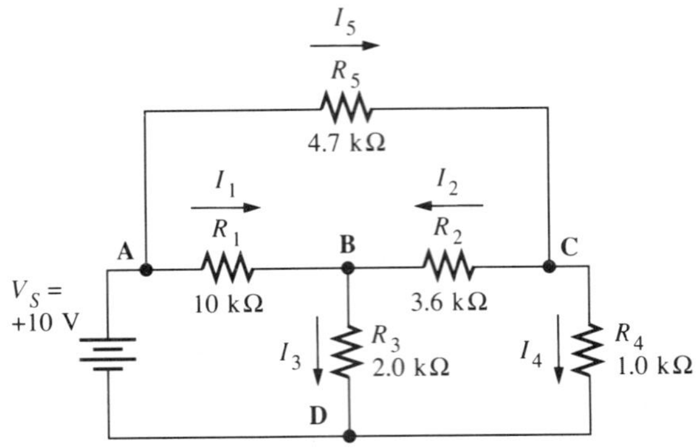

Figure 15–4

FOR FURTHER INVESTIGATION:

Mesh and node equations are useful for solving problems with more than one source without requiring the superposition theorem. Refer to Experiment 11 (superposition) and write loop equations for the two "windows" for the circuit shown in Figure 11–1. (A window is an open section of a circuit with no crossovers. It is bounded on all sides by sources, components, or wiring.) Solve for the currents in each resistor, then apply Ohm's law and compute the voltage drop across each resistor. Compare your loop equation results with the measured voltages tabulated in Table 11–2.

APPLICATION PROBLEM:

An interesting application of nodal analysis is found in developing the equations for a Howland current source. In this application problem, you will not actually construct the circuit but rather set up the equations to solve for the output current in terms of the input voltage. The circuit uses an operational amplifier (op-amp), a high-gain dc amplifier available in integrated circuit form. By using two approximations for the ideal op-amp, the equations for the current source can be found. The first approximation is that no current flows into the inputs of the op-amp; the second is that the voltage difference across the input terminals is zero.

The circuit is shown in Figure 15–5(a). All resistors labeled R are equal. To begin, notice that the upper resistance path consists of a simple unloaded voltage divider (because of the first approximation). Therefore, the voltage on the inverting input terminal is one-half of the output voltage, V_{out}. The second approximation means that this voltage is also found on the noninverting side of the op-amp. With these approximations, we can redraw the resistors around node **A,** as shown in Figure 15–5(b). Write a node equation for point **A** and prove that I_L is equal to V_{in}/R.

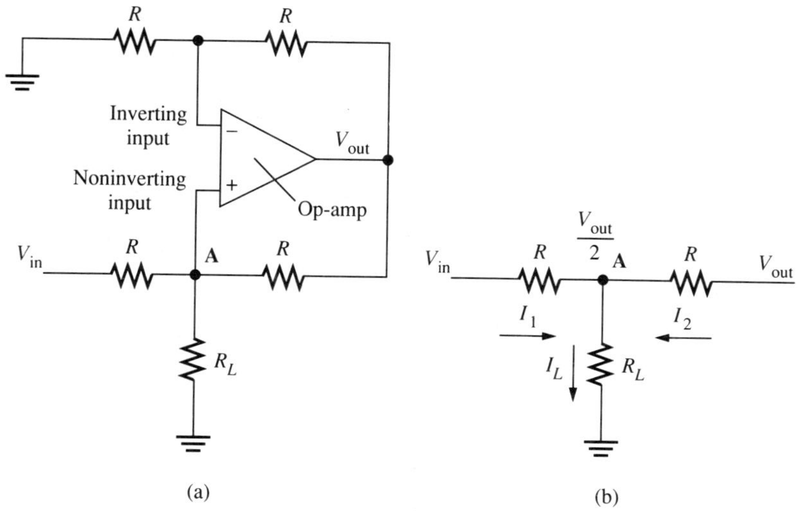

R
R
Inverting
input
V_{out}
Noninverting
input
Op-amp
V_{in}
R
A
R
R_L
(a)
$\frac{V_{out}}{2}$
V_{in}
R
A
R
V_{out}
I_1
I_2
I_L
R_L
(b)

Figure 15–5

Report for Experiment 15

Name ____________
Date ____________
Class ____________

OBJECTIVES:

PROCEDURE:

DATA:

Loop Equations:

Table 15–1

Component	Listed Value	Measured Value
R_1	10 kΩ	
R_2	3.6 kΩ	
R_3	2.0 kΩ	
R_5	1.0 kΩ	
R_5	4.7 kΩ	

Table 15–2

	Computed Current
I_A	
I_B	
I_C	
I_1	
I_2	
I_3	
I_4	
I_5	

Table 15–3

	Computed	Measured
V_1		
V_2		
V_3		
V_4		
V_5		

Node Equations:

RESULTS AND CONCLUSIONS:

FURTHER INVESTIGATION RESULTS:

APPLICATION PROBLEM RESULTS:

EVALUATION AND REVIEW QUESTIONS:

1. As a consistency check on your results, mentally remove R_4 from the bridged-T circuit and compute a Thevenin equivalent for the remaining circuit. Then compute the current in R_4 if it were connected to your Thevenin equivalent and show that you obtain the same I_4.

2. The circuit shown in Figure 15–6(a) shows a bridged-T circuit with all resistances equal to the Thevenin driving resistance of 1.0 kΩ. Assume you need to find the equivalent Thevenin driving impedance of the entire circuit looking from R_L. A useful "trick" is to replace the Thevenin source with its equivalent resistance and add a 1 V source on the output, as shown in Figure 15–6(b). The current leaving the 1 V source can now be calculated by setting up loop equations. Prove that this current is equal to 1 mA (showing that the bridged-T did not change the original driving impedance!).

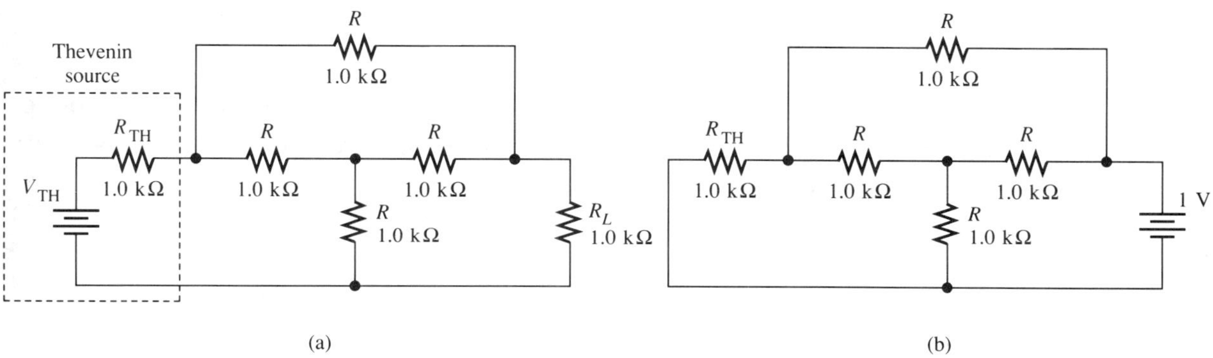

Figure 15–6

3. Suppose the loop currents in this experiment had been drawn in the opposite direction (counterclockwise). What effect does this have on the current in each resistor?

4. The ratio of output voltage to input voltage for a network is called the transfer function of the network. Find the transfer function for the network shown in Figure 15–7 by solving the loop equations (written in the box) for the paths shown. All resistors are equal. Notice that V_{out} is given by RI_B. Then solve for the ratio of V_{out}/V_{in}.

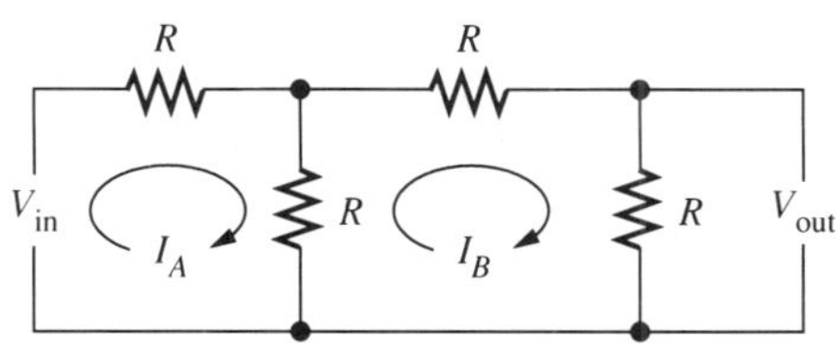

$$2RI_A - RI_B = V_{in}$$

$$-RI_A + 3RI_B = 0$$

Figure 15–7

5. Write the node equations for the Wheatstone bridge circuit shown in Figure 13–1(a).

6. (a) What type of equations result from an application of Kirchhoff's voltage law?

(b) What type of equations result from an application of Kirchhoff's current law?

TECH TIP REVIEW:

Both of the following problems relate to the TECH TIP circuit shown in Figure 9–10 of the Floyd text. Assume the circuit is not operating properly. You have made certain measurements; from these measurements, can you determine the most probable cause for each of the following problems?

1. A measurement of the voltages on the output terminal board indicates that the voltage measured with respect to ground is zero; however, a measurement of the voltage drop *across* each resistor is normal.

2. The positive output voltage is +8.32 V and the negative output voltage is −7.38 V with the loads connected.

16 Magnetic Devices

OBJECTIVES:

After performing this experiment, you will be able to:

1. Determine the pull-in voltage and release voltage for a relay.
2. Connect relay circuits including a relay latching circuit.
3. Explain the meaning of common relay terminology.

READING:

Floyd, *Principles of Electric Circuits,* Sections 10–1 to 10–7

MATERIALS NEEDED:

One DPDT relay; 6 V dc or 12 V dc coil
Two LEDs; one red, one green
One SPST switch
Two 330 Ω resistors
For Further Investigation: One CdS photocell (Radio Shack 276-116 or equivalent)

SUMMARY OF THEORY:

Magnetism plays an important role in a number of electronic components and devices. These include inductors, transformers, relays, solenoids, transducers, magnetic tape, meters, and many other devices used in the physical sciences. Magnetic fields are associated with the movement of electric charges, a fact discovered in 1820 by Hans Christian Oersted. By forming a coil, the magnetic field lines could be concentrated, a concept that is used in most devices that use the electromagnetic field. Wrapping the coil on a magnetic core material such as iron, silicon steel, or permalloy provides two additional advantages. First, the magnetic flux is increased because the *permeability* of these materials is much higher than air. Permeability is a measure of how easily magnetic field lines pass through a material. Permeability is not a constant for a material but depends on the amount of flux in the material. The second advantage of using a magnetic core material is that the flux is more concentrated.

A common electromagnetic device is the *relay.* The relay is an electromagnetic switch with one or more sets of contacts used for a number of switching applications including controlling large currents or voltages. The contacts are controlled by an electromagnet—a coil of wire wrapped on a magnetic material—that can be designed for either ac or dc operation. Contacts are specified as either *normally open* (NO) or *normally closed* (NC) when no voltage is applied to the coil. Relays are *energized* by applying the rated voltage to the coil. This causes the contacts to either close or open.

Relays, like mechanical switches, are specified in terms of the number of independent switches (called *poles*) and the number of contacts (called *throws*). Thus, a single-pole, double-throw (SPDT) relay has a single switch with two contacts—one normally open (NO) and one normally closed (NC). An example of such a relay in a circuit is shown in Figure 16–1(a). With S_1 open, the motor is off and the light is on. When S_1 is closed, coil CR_1 is energized, causing the NO contacts to close and the NC contacts to open. This applies line voltage to the motor and at the same time removes line voltage from the light. Figure 16–1(a) is drawn in a manner similar to many industrial schematics—sometimes referred to as a ladder schematic. The NC contacts are indicated on this schematic with a diagonal line drawn through them. An alternative way of drawing the same schematic is shown in Figure 16–1(b). In this drawing, the relay contacts are drawn as a switch.

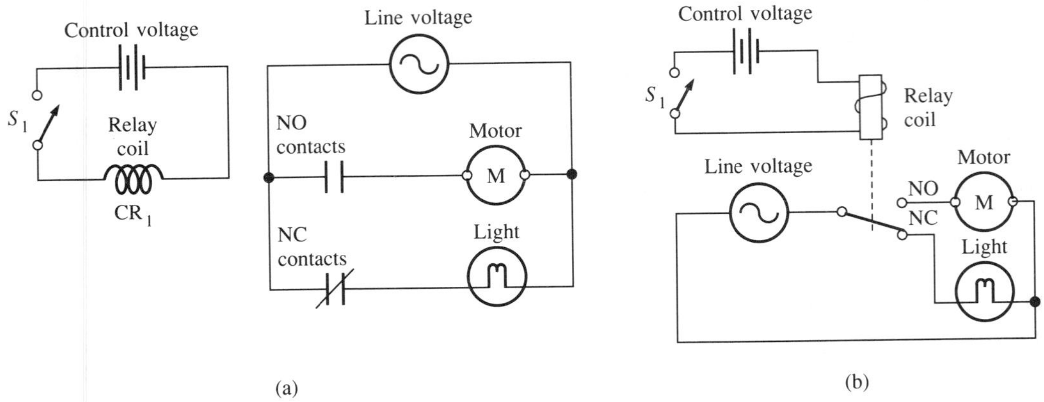

Figure 16–1

Manufacturers specify relays in terms of the ratings for the coil voltage and current, maximum contact current, operating time, and so forth. Manufacturer's specifications include the various ratings and show the location of contacts and coil. Frequently, if these are not available, the technician can determine the electrical wiring of contacts and coil by inspection and ohmmeter tests.

PROCEDURE:

1. Obtain a double-pole, double-throw (DPDT) relay with either a 6 V dc or 12 V dc coil. The terminals should be numbered. Inspect the relay and try to determine which terminals are connected to the coil and which are connected to the contacts. The connection diagram is frequently drawn on the relay. Check the coil with an ohmmeter. It should indicate the coil resistance. Check contacts with the ohmmeter. NC contacts should read near 0 Ω, whereas NO contacts should read infinite resistance. You may have difficulty determining which contact is movable until the coil is energized. In the space provided in the report, draw a diagram of your relay, showing the coil, all contacts, and terminal numbers. (Refer to Figure 16–1 for an example.) Record the coil resistance on your drawing.

2. Connect the circuit shown in Figure 16–2, which is drawn as a ladder diagram. In this circuit, only one pole of the relay is used. The movable arm is connected to the negative side of V_{S2}. Note carefully the direction of the light-emitting diodes (LEDs). LEDs are polarized and must be connected in the correct direction. V_{S1} is the control voltage and should be set to the specified coil voltage for the relay. V_{S2} represents a line voltage that is being controlled. For safety, a low voltage is used. Set V_{S2} for 5.0 V. If the circuit is correctly connected, the green LED should be on with S_1 open. Close S_1 and verify that the red LED turns on and the green LED goes off.

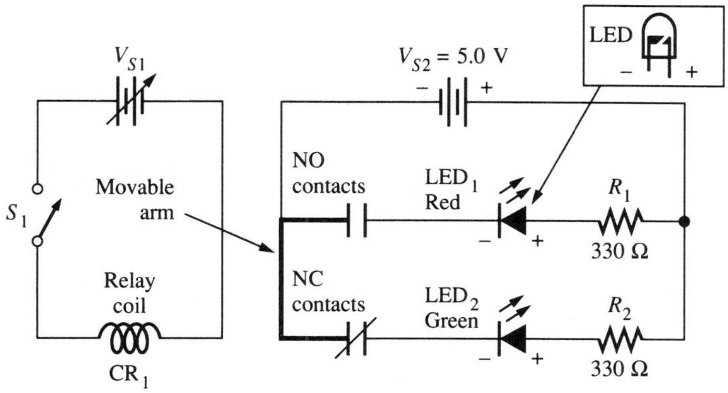

Figure 16–2

3. In this step, you will determine the *pull-in voltage* of the relay. The *pull-in voltage* is the minimum value of coil voltage which will cause the relay to switch. Turn V_{S1} to its lowest setting. With S_1 closed, gradually raise the voltage until the relay trips as indicated with the light-emitting diodes (LEDs). Record the pull-in voltage in Table 16–1.

4. The *release voltage* is the value of the coil voltage at which the contacts return to the unenergized position. Gradually lower the voltage until the relay resets to the unenergized position, as indicated by the LEDs. Record the release voltage in Table 16–1.

5. Repeat steps 3 and 4 for three trials, entering the results of each trial in Table 16–1.

6. Compute the average pull-in voltage and the average release voltage. Enter the averages in Table 16–1.

7. In this step you will learn how to construct a latching relay. Connect the unused NO contacts from the other pole on the relay in parallel with S_1, as illustrated in Figure 16–3. Set V_{S1} for the rated coil voltage. Close and open S_1. Describe your observations in the space provided in the report.

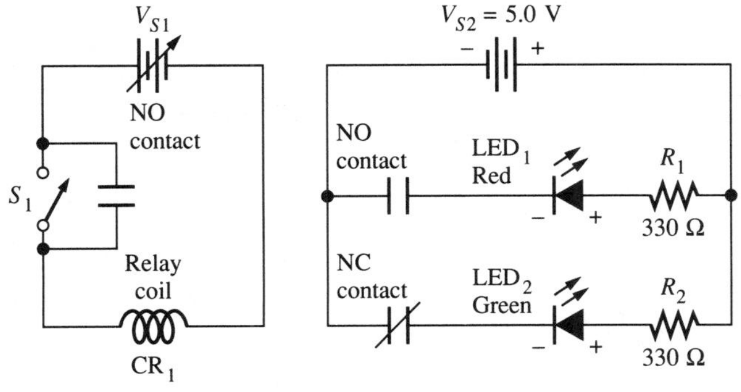

Figure 16–3

8. Remove the NO contact from around S_1. Connect the NC contact in *series* with S_1, as shown in Figure 16–4. Explain what happens in your report.

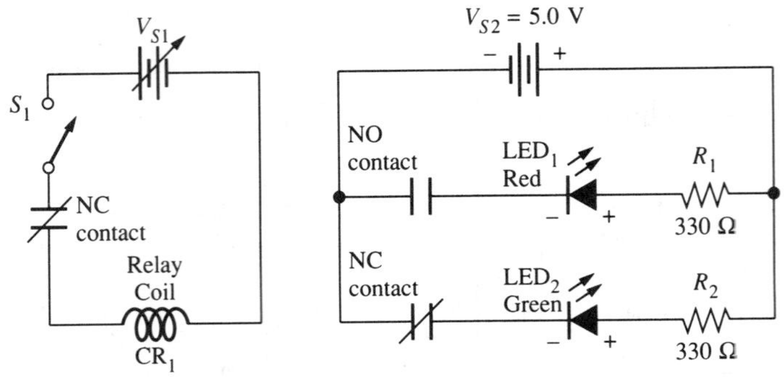

Figure 16–4

FOR FURTHER INVESTIGATION:
A cadmium sulfide (CdS) photocell is a device that changes its resistance when light strikes it. Devise a circuit in which a CdS cell controls the energizing of a relay. Build and test your circuit. Summarize your results in your report. Include the schematic, the measurements you made, and conclusions about your design.

APPLICATION PROBLEM:
This application problem requires the analysis and design modification of a control system for a commercial electric garage-door control, as shown in Figure 16–5. There are two separate relays with an NO and NC contact shown for each relay. The first number shown under the relay contacts represents the relay number and the second number shown represents the relay contact numbers. The letters CR are used for the coil. The door is controlled with three manual push buttons labeled OPEN, CLOSE, and STOP. In this problem, you need to analyze the operation of the circuit by answering the questions listed in the report. After you understand its operation, draw a new schematic showing how you could modify the circuit to add another complete set of push buttons (OPEN, CLOSE, and STOP) to allow controlling the door from a second point.

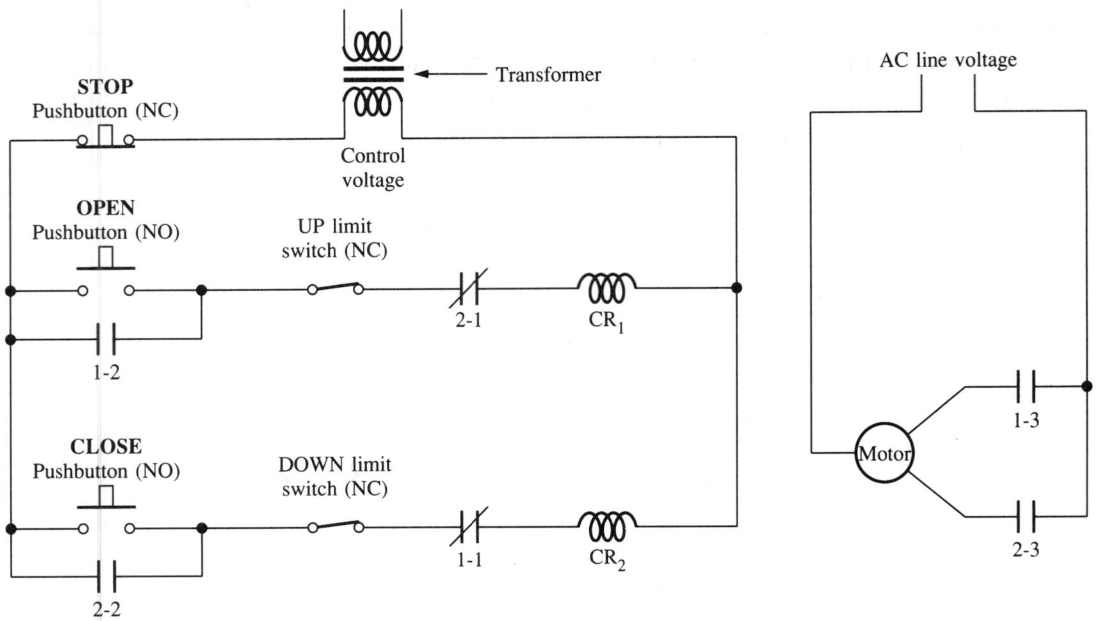

Figure 16–5

Report for Experiment 16

Name ______________
Date ______________
Class ______________

OBJECTIVES:

PROCEDURE:

DATA:

Step 1 (Relay Drawing):

Table 16–1

	Pull-in Voltage	Release Voltage
Trial 1		
Trial 2		
Trial 3		
Average		

Step 7 (Latching Relay Observations):

Step 8 (Figure 16–4 Observations):

RESULTS AND CONCLUSION:

FURTHER INVESTIGATION RESULTS:

APPLICATION PROBLEM RESULTS:

Questions for the Application Problem:

1. When the OPEN pushbutton is *momentarily* closed and released, the door continues to open. What accounts for this action?

2. What normally stops the action of opening the door?

3. **Assume the STOP pushbutton is momentarily depressed while the door is closing: what action takes place? Why doesn't the door resume its motion when the STOP pushbutton is released?**

4. Why are the contacts for relay CR_1 in series with the coil of relay CR_2?

5. Which contacts are responsible for applying power to the motor for closing the door?

Schematic for Controlling Door from a Second Point:

EVALUATION AND REVIEW QUESTIONS:

1. Using the average *pull-in voltage* and the measured resistance of your relay coil, compute the average pull-in current. The pull-in current is defined as the minimum value of coil current at which the switching function is completed.

2. Repeat question 1 for the *release voltage* using the average of the measured values.

3. Hysteresis is a property of a system when its response is dependent on the immediate past state of the system. For a relay, the pull-in and release voltage are not the same due to hysteresis. The hysteresis of a relay can be defined as the difference between the pull-in and the release voltage. Compute the hysteresis of your relay.

4. (a) Explain the difference between (a) SPDT and (b) DPST.

   (b) Explain the meaning of NO and NC.

5. For the circuit of Figure 16–1, assume that when S_1 is closed, the light stays on and the motor remains off.

   (a) Name two possible faults that could account for this.

   (b) What procedure would you suggest to isolate the fault?

6. Name two advantages of using a magnetic-core material in a relay coil.

TECH TIP REVIEW:
Each of the following problems relate to the TECH TIP circuit shown in Figure 10–39 of the Floyd text. Assume the alarm system needs to be expanded to include the following features. Show with a drawing how you would implement each new feature.

1. Red and a green indicator lights are needed to indicate if the system is ON or OFF. The indicators are to be made using light-emitting diodes (LEDs) and will operate from the +9 V battery with a series 3.3 kΩ resistor. LEDs allow current to flow in only one direction and are drawn as shown in Figure 16–4, except larger resistors are needed due to the larger supply voltage. The green LED should be ON when the system ON/OFF switch is open (system OFF). The red indicator light should be ON when the system switch is closed (system ON).

2. A system TEST push button is to be added. The system TEST is a momentary contact single-pole, double-throw (SPDT) push-button switch that is normally closed (NC). When the push-button is pressed, the relay is activated and the house light turns ON but the alarm does *not* sound and the holding contacts of the relay are *not* active.

17 The Oscilloscope

OBJECTIVES:

After performing this experiment, you will be able to:

1. Explain the four major groups of controls on the oscilloscope.
2. Use an oscilloscope to measure ac and dc voltages.

READING:

Floyd, *Principles of Electric Circuits,* Sections 11–1 through 11–7 and 11–9
Reference Guide to Laboratory Instruments, p. 9

MATERIALS NEEDED:

None

SUMMARY OF THEORY:

The oscilloscope is an extremely versatile instrument that lets you see a picture of the voltage in a circuit as a function of time. The voltage is converted into a visible display by a cathode-ray tube (CRT), a vacuum device similar to a television picture tube. The oscilloscope can be broken into four functional blocks, as illustrated in Figure 17–1. The input signal is coupled through a vertical section to a display section, causing the beam to move up and down as the input voltage moves up or down. A trigger section sends a signal to the horizontal section causing the beam to move (or *sweep*) across the screen. This horizontal movement must be synchronized with the vertical signal to present a stable pattern on the screen.

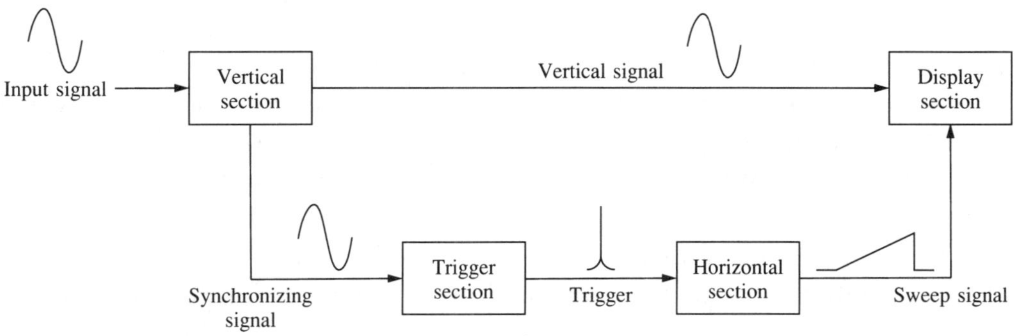

Figure 17–1

Controls for each of the functional blocks are usually grouped together. Frequently, there are color clues to help you identify groups of controls. Look for the controls for each functional group on your oscilloscope. The display controls include INTENSITY, FOCUS, and BEAM FINDER. The vertical controls include input COUPLING, VOLTS/DIV, vertical POSITION, and channel selection (CH1, CH2, DUAL, ALT, CHOP). The triggering controls include MODE, SOURCE, trigger COUPLING, trigger LEVEL, and others. The horizontal controls include the SEC/DIV, MAGNIFIER, and horizontal POSITION controls. Details of these controls are explained in the referenced reading and in the operator's manual for the oscilloscope.

With all the controls to learn, you may experience difficulty obtaining a trace. If you do not see a trace, start by setting the SEC/DIV control to 0.1 ms/div, select AUTO triggering, select CH1, and press the BEAM FINDER. Keep the BEAM FINDER button depressed and use the vertical and horizontal POSITION controls to center the trace. If you still have trouble, check the INTENSITY control.

Because the oscilloscope can show a voltage-versus-time presentation, it is easy to make ac voltage measurements with a scope. However, care must be taken to equate these measurements with meter readings. Typical digital multimeters show the *rms* (root-mean-square) value of a sinusoidal waveform. This value represents the effective value of an ac waveform when compared to a dc voltage when both produce the same heat in a given load. Usually the *peak-to-peak* value is easiest to read on an oscilloscope. The relationship between the ac waveform as viewed on the oscilloscope and the equivalent rms reading that a DMM will give is illustrated in Figure 17–2.

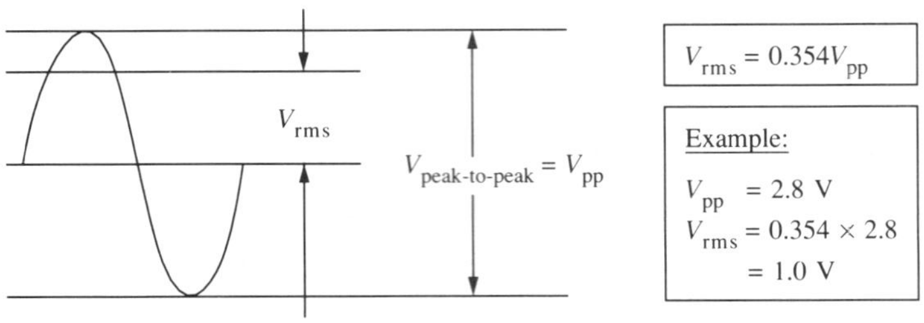

Figure 17–2

Waveforms that are not sinusoidal cannot be directly compared with an oscilloscope and DMM except for the dc component. (See Application Problem.) The dc level of any waveform can be represented by a horizontal line which splits the waveform into equal areas above and below the line. For a sinusoidal wave, the dc level is always halfway between the maximum and minimum excursions. The dc component can be correctly read by a DMM no matter what the shape of the wave when it is in the DC volts mode.

The amplitude of any periodic waveform can be expressed in one of four ways: the peak-to-peak, the peak, the rms, or the average value. The peak-to-peak value of any waveform is the total magnitude of the change and is *independent* of the zero position. The peak value is the maximum excursion of the wave and is usually referenced to the dc level of the wave. If you want to indicate that the reported value includes a dc offset, you need to make this clear by stating both the maximum and minimum excursions of the waveform.

An important part of any oscilloscope measurement is the oscilloscope probe. The type of probe that is generally furnished with an oscilloscope by the manufacturer is called an *attenuator probe* because it attenuates the input by a known factor. The most common attenuator probe is the $\times 10$ probe, because it reduces the input signal by a factor of 10. It is a good idea, before making any measurement, to check that the probe is properly compensated, meaning that the frequency response of the probe/scope system is flat. Probes have a small variable capacitor either in the probe tip or a small box that is part of the input connector. This capacitor is adjusted while observing a square wave to ensure that the displayed waveform has vertical sides and square corners. Most oscilloscopes have the square-wave generator built in for the purpose of compensating the probe.

PROCEDURE:

1. Review the front panel controls in each of the major groups. Then turn on the oscilloscope, select CH1, set the SEC/DIV to 0.1 ms/div, select AUTO triggering, and obtain a line across the face of the CRT.

2. Turn on your power supply and use the DMM to set the output for 1.0 V. Now we will use the oscilloscope to measure this dc voltage from the power supply. The following steps will guide you:
   (a) Place the vertical COUPLING (AC-GND-DC) in the GND position. This disconnects the input to the oscilloscope. Use the vertical POSITION control to set the ground reference level on a convenient graticule line near the bottom of the screen.
   (b) Set the CH1 VOLTS/DIV control to 0.2 V/div. Check that the red vernier control is in the CAL position or your measurement will not be accurate.
   (c) Place the oscilloscope probe on the positive side of the power supply. Place the oscilloscope ground on the power supply common. Move the vertical coupling to the DC position. The line should jump up on the screen by 5 divisions. *Note that* 5 *divisions times* 0.2 V *per division is equal to* 1.0 V *(the supply voltage).* Multiplication of the number of divisions of deflection times volts per division is equal to the voltage measurement.

3. Set the power supply to each voltage listed in Table 17–2. Measure each voltage using the above steps as a guide. The first line of the table has been completed as an example. To obtain accurate readings with the oscilloscope, it is necessary to select the VOLTS/DIV that gives several divisions of change between the ground reference and the voltage to be measured. The readings on the oscilloscope and meter should agree with each other within approximately 3%.

4. Before viewing ac signals, it is a good idea to check the probe compensation for your oscilloscope. To check the probe compensation, set the VOLTS/DIV control to 0.1 V/div, the AC-GND-DC coupling control to DC, and the SEC/DIV control to 2 ms/div. Touch the probe tip to the PROBE COMP connector. You should observe a square wave with a flat top and square corners. If necessary, adjust the compensation to achieve a good square wave.

5. Set the function generator for an ac waveform with a frequency of 1.0 kHz. Adjust the amplitude of the function generator for 1.0 V_{rms} as read on your DMM. Set the SEC/DIV control to 0.2 ms/div and the VOLTS/DIV to 0.5 V/div. Connect the scope probe and its ground to the function generator. Adjust the vertical POSITION control and the trigger LEVEL control for a stable display near the center of the screen. You should observe approximately two cycles of an ac waveform with a peak-to-peak amplitude of 2.8 V. This represents 1.0 V_{rms}, as shown in Figure 17–2.

6. Use the DMM to set the function generator amplitude to each value listed in Table 17–3. Repeat the ac voltage measurement as outlined in step 4. The first line of the table has been completed as an example. Remember, to obain accurate readings with the

oscilloscope, you should select a VOLTS/DIV setting that gives several divisions of deflection on the screen.

FOR FURTHER INVESTIGATION:

Most function generators have a control that allows you to add or subtract a dc offset voltage to the signal. Set up the function generator for a 1.0 kHz sine-wave signal, as shown in Figure 17–3. To do this, the AC-GND-DC coupling switch oscilloscope should be in the DC position and the offset control should be adjusted on the function generator. When you have the signal displayed on the oscilloscope face, switch the AC-GND-DC coupling switch into the AC position. Explain what this control does. Then measure the signal with your DMM. First measure it in the AC VOLTAGE position; then measure in the DC VOLTAGE position. How does this control differ from the AC-GND-DC coupling switch on the oscilloscope? Summarize your findings.

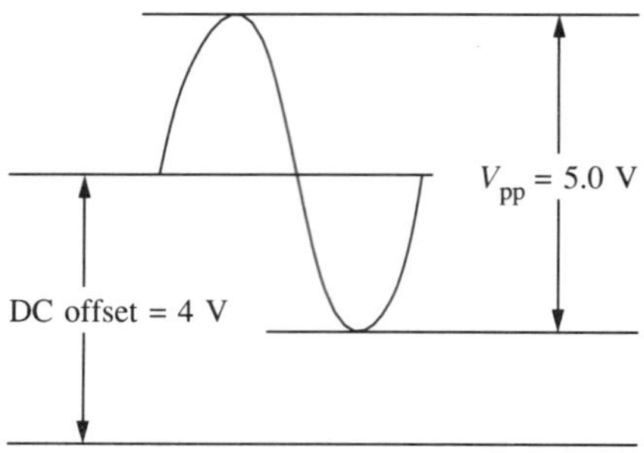

Figure 17–3

APPLICATION PROBLEM:

Suppose you want to compare the reading of an oscilloscope to that of a DMM or a VOM when the waveform is not sinusoidal. Most meters do not read the true rms value of a waveform. Instead, they read the absolute average value and convert the reading to an equivalent rms value *based on a sinusoidal input signal.* When the input waveform is not sinusoidal, the meter reading is in error. The meter can correctly read the dc offset of any waveform, but in the ac position it is calibrated only for a sinusoidal input. All average reading meters (including most DMMs and VOMs) apply a form factor (FF) to the *absolute magnitude* of the average value of a sine wave to obtain the value shown on the display. By knowing the FF, you can compute the expected reading of a nonsinusoidal waveform on the DMM or VOM and compare it with the oscilloscope.

The FF used by meters is the ratio of the true rms value to the absolute average value of a sinusoidal wave. This ratio is 1.11 for full-wave meters.* Thus a 10 V_{rms} reading is sensed by a full-wave reading meter as 9.00 V_{avg} and displayed as $1.11 \times 9.00 = 10.0$ V_{rms}. For a nonsinusoidal waveform, the DMM will read the average value of the waveform and *incorrectly* multiply it by the sinusoidal FF. By dividing the displayed reading by the FF, you can obtain the correct *average* value for the waveform on the display. The rms or peak value can then be found by multiplying the computed average value by the appropriate conversion factor for that waveform. To compare the oscilloscope reading to that of the DMM, you will need to convert from peak-to-peak to average for that waveform and then multiply by the FF

---

*If your meter is a half-wave meter, the form factor is 2.22.

for your meter. The FF and ratios of peak, avg, and rms voltages for several waveforms are listed in Table 17–1.

Using the oscilloscope, set up a 1.0 V_p (2.0 V_{pp}) waveform at 100 Hz from your function generator for each waveform shown. Use AC coupling on your oscilloscope and on your DMM. From the conversion factors listed in Table 17–1, calculate the average ac voltage for the waveform you are viewing. Then compute the reading that you expect on your DMM by multiplying by the FF for your meter. Tabulate your results in Table 17–4 of your report. The last two columns of Table 17–4 should agree within experimental error.

Table 17–1

Waveform		$FF = \frac{V_{rms}}{V_{avg}}$	Peak, average, and rms values		
			V_p	V_{avg}	V_{rms}
Sine	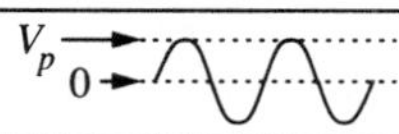	1.11	1.00	0.637	0.707
Square	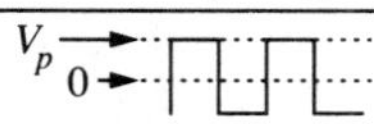	1.00	1.00	1.00	1.00
Triangle	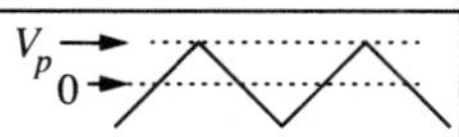	1.15	1.00	0.500	0.576
Sawtooth	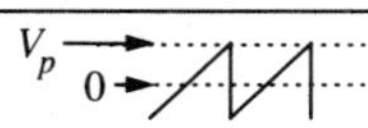	1.15	1.00	0.500	0.576

Report for Experiment 17

Name ______________
Date ______________
Class ______________

OBJECTIVES:

PROCEDURE:

DATA:

Table 17–2

Power Supply Setting	VOLTS/DIV Setting	Number of Divisions of Deflection	Oscilloscope (measured voltage)	DMM (measured voltage)
1.0 V	0.2 V/div	5.0 DIV	1.0 V	1.0 V
2.5 V				
4.5 V				
8.3 V				

Table 17–3

Function Generator Amplitude	VOLTS/DIV Setting	Number of Divisions (peak-to-peak)	Oscilloscope measured (peak-to-peak)	Oscilloscope measured (rms)
1.0 V_{rms}	0.5 V/div	5.6 div	2.8 V_{pp}	1.0 V_{rms}
2.2 V_{rms}				
3.7 V_{rms}				
4.8 V_{rms}				

RESULTS AND CONCLUSION:

FURTHER INVESTIGATION RESULTS:

APPLICATION PROBLEM RESULTS:

Table 17–4

	Oscilloscope Readings			
Waveform	V_p (Measured)	V_{avg} (Computed)	$V_{avg} \times FF =$	DMM Reading
Sine				
Square				
Triangle				
Sawtooth				

EVALUATION AND REVIEW QUESTIONS:

1. (a) Compute the percent difference between the DMM measurement and the oscilloscope measurement for each dc voltage measurement summarized in Table 17–2.

   (b) Which do you think is most accurate? ____________________Why?

2. Briefly, describe the four major groups of controls on the oscilloscope and the purpose of each group.

3. If you are having difficulty obtaining a stable display, which group of controls should you adjust?

4. (a) If an ac waveform has 3.4 divisions from peak-to-peak and the VOLTS/DIV control is set to 5.0 V/div, what is the peak-to-peak voltage?

   (b) What is the rms voltage?

5. If you want to view the amplitude of an ac waveform that is 20.0 V_{rms}, what setting of the VOLTS/DIV control is best?

6. The most accurate way to measure a waveform on an oscilloscope is to use a large portion of the display area. Why?

18 Sine-Wave Measurements

OBJECTIVES:

After performing this experiment, you will be able to:

1. Measure the period and frequency of a sine wave using an oscilloscope.
2. Measure across ungrounded components using the difference function of an oscilloscope.

READING:

Floyd, *Principles of Electric Circuits,* Sections 11–1 through 11–7 and 11–9
Reference Guide to Laboratory Instruments, p. 9

MATERIALS NEEDED:

One 2.7 kΩ resistor, one 6.8 kΩ resistor

SUMMARY OF THEORY:

Imagine a weight suspended from a spring. If you stretch the spring and then release it, it will bob up and down with a regular motion. The distance from the rest point to the highest (or lowest) point is called the *amplitude* of the motion. As the weight moves up and down, the time for one complete cycle is called a *period* and the number of cycles it moves in a second is called the *frequency.* This cyclic motion is called *simple harmonic motion.* A graph of simple harmonic motion as a function of time produces a sine wave, the most fundamental waveform in nature. It is generated as the natural waveform from an ac generator. Figure 18–1 illustrates these definitions.

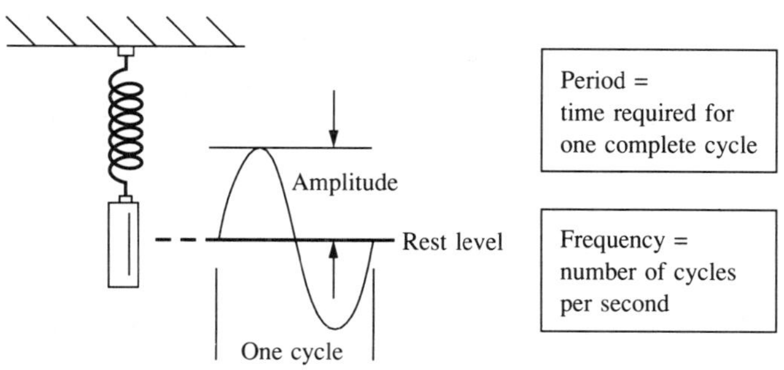

Figure 18–1

Sine waves can also be generated from uniform circular motion. Imagine a circle turning at a constant rate. The *projection* of the endpoint of the radius vector moves with simple harmonic motion. If the endpoint is plotted along the x-axis, the resulting curve is a sine wave, as illustrated in Floyd's text. This method is frequently used to show the phase relationship between two sine waves of the same frequency.

The sine wave has another interesting property. Different sine waves can be added together to give new waveforms. In fact, any repeating waveform such as a ramp or square wave can be made up of a group of sine waves. This property is useful in the study of the response of circuits to various waveforms.

The oscilloscope is a powerful tool for viewing any waveform in a circuit. In this experiment, you will make time measurements on sine waves. To measure time, the oscilloscope generates an internal sawtooth waveform that is applied to the horizontal deflection plates. This signal is the *sweep,* or time base, of the oscilloscope and is controlled by the SEC/DIV control on the front panel of the oscilloscope. The control generally has a vernier and a magnifier associated with it. It is important that these controls are in the calibrated positions whenever a time measurement is made.

PROCEDURE:

1. Set the function generator for a 1.0 V_{pp} sine wave at a frequency of 1.25 kHz. Then set the oscilloscope SEC/DIV control to 0.1 ms/div in order to show one complete cycle on the screen. *The expected time for one cycle (the period) is the reciprocal of* 1.25 kHz, *which is* 0.8 ms. With the SEC/DIV control at 0.1 ms/div, one cycle requires 8.0 divisions across the screen. This information is presented as an example on line 1 of Table 18–1.

2. Change the function generator to each frequency listed in Table 18–1. Complete the table by computing the expected period and then measuring the period on the oscilloscope.

3. In this step you will need to use a two-channel oscilloscope with two probes, one connected to each channel. Frequently, a voltage measurement is needed across an ungrounded component. If the oscilloscope ground is at the same potential as the circuit ground, then the process of connecting the probe will put an undesired ground path in the circuit. Figure 18–2 illustrates this.

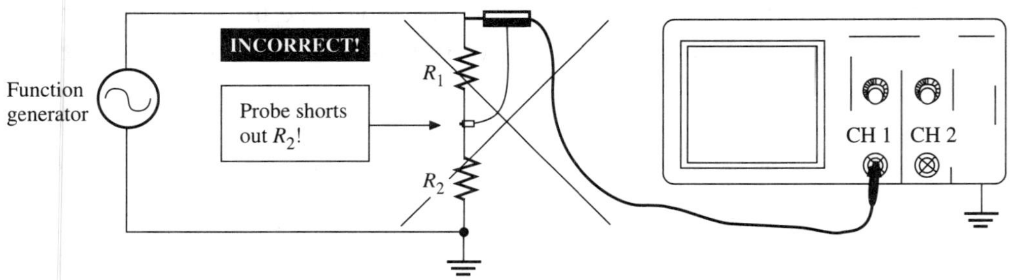

Figure 18–2

The correct way to measure the voltage across the ungrounded component is to use two channels and select the subtract mode—sometimes called the *difference function,* as illustrated in Figure 18–3. The difference function subtracts the voltage measured on channel 2 from the voltage measured on channel 1. It is important that both channels have the same vertical sensitivity—that is, that the VOLTS/DIV setting is the same on both channels and they are both calibrated.

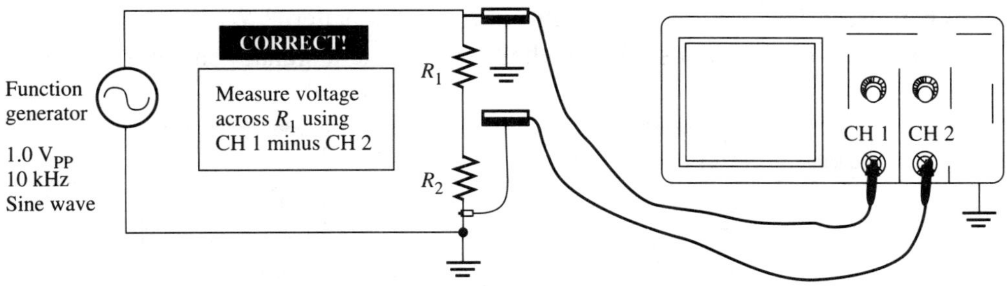

Figure 18–3

Connect the circuit shown in Figure 18–3. Use a 2.7 kΩ resistor for R_1 and a 6.8 kΩ resistor for R_2. Set the function generator for a 1.0 V_{pp} sine wave at 10 kHz. Channel 1 will show the voltage from the generator. Channel 2 will show the voltage across R_2. The difference (CH1 subtract CH2) will show the voltage across R_1. Some oscilloscopes require that you ADD the channels and INVERT channel 2 in order to measure the difference in the signals.* Complete Table 18–2 for the voltage measurements. Use the voltage divider rule to check that your measured voltages are reasonable.

FOR FURTHER INVESTIGATION:

It is relatively easy to obtain a stable display on the oscilloscope at higher frequencies. It is more difficult to obtain a stable display with slower signals, especially those with very small amplitude. Set the signal generator on a frequency of 5 Hz. Try to obtain a stable display. You will probably have to use NORMAL triggering and carefully adjust the trigger LEVEL control. After you obtain a stable display, try turning the amplitude of the function generator to its lowest setting. Can you still obtain a stable display? Measure the lowest amplitude signal for which you obtain a stable trace.

APPLICATION PROBLEM:

Some interesting and useful patterns called Lissajous figures can be used to compare two frequencies that are an exact multiple of each other. Start by setting two separate function generators to a 1.0 V_{pp} sine wave at 1.0 kHz, as indicated on the generator dials. Then connect the circuit shown in Figure 18–4. Do not put the oscilloscope into X-Y mode until you have a signal connected from each generator (to avoid leaving a dot on the screen). Place both the CH1 and CH2 vertical amplifiers on 0.2 V/div.

Observe the pattern on the oscilloscope. Try to stop it by carefully adjusting the frequency dial of the second function generator. After observing the pattern, change the frequency of the second generator to 2.0 kHz and observe the new pattern. Again try to stop the figure by carefully adjusting the second generator's frequency. Notice the number of points at which the pattern touches the *x*-axis and the number of points at which the pattern touches the *y*-axis. Move the second generator to 3 kHz and note the effect. Try adjusting the amplitudes of the generators. Describe your observations in your

---

*If you do not have difference channel capability, then temporarily reverse the components to put R_1 at circuit ground. This can be easily accomplished with a lab breadboard but is usually not practical in a manufactured circuit. Although it is possible to isolate the oscilloscope ground and then use one channel to make the measurement, the procedure is not recommended.

report. What effect does the frequency ratio have on the pattern? What effect does the amplitude control have? How could you use this method to calibrate a signal generator against a standard?

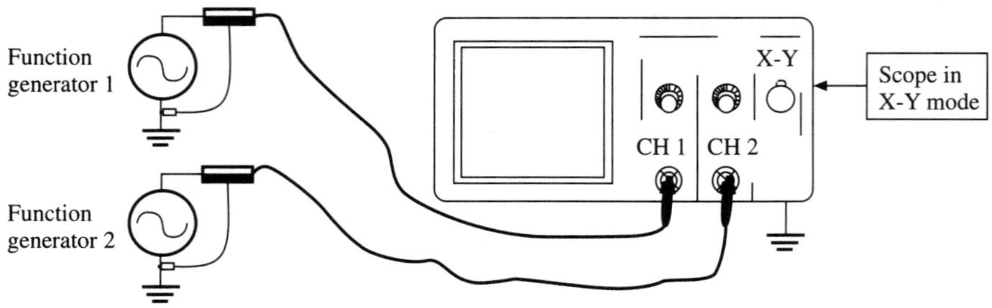

Figure 18–4

Report for Experiment 18

Name ____________________
Date ____________________
Class ____________________

OBJECTIVES:

PROCEDURE:

DATA:

Table 18–1

Function Generator Dial Frequency	Computed Period	Oscilloscope SEC/DIV	Number of Divisions	Measured Period
1.25 kHz	0.8 ms	0.1 ms/div	8.0 div	0.8 ms
1.90 kHz				
24.5 kHz				
83.0 kHz				
600.0 kHz				

Table 18–2

	Function Gen. Voltage	Voltage across R_1	Voltage across R_2
measured			
computed	1.0 V_{pp}		

RESULTS AND CONCLUSION:

FURTHER INVESTIGATION RESULTS:

APPLICATION PROBLEM RESULTS:

EVALUATION AND REVIEW QUESTIONS:

1. (a) Compare the computed and measured periods for the sine waves in Table 18–1. Calculate the percent difference for each row of the table.

   (b) What measurement errors account for the percent differences?

2. Using the measured voltages in Table 18–2, show that Kirchhoff's voltage law is satisfied.

3. An oscilloscope display shows one complete cycle of a sine wave in 6.3 divisions. The SEC/DIV control is set to 20 ms/div.
   (a) What is the period? ____________
   (b) What is the frequency? ____________

4. You wish to display a 10 kHz sine wave on the oscilloscope. What setting of the SEC/DIV control will show one complete cycle in 10 divisions?

   SEC/DIV = ____________

5. Explain how to measure the voltage across an ungrounded component.

6. Explain when to select CHOP or ALTernate when viewing two signals on an oscilloscope.

TECH TIP REVIEW:
Each of the following questions relate to the TECH TIP circuit shown in Figure 11–61 of the Floyd text.

1. Assume you want to observe the low-level signal on the *input* to the audio preamplifier (marked point 4) with your oscilloscope but you are having difficulty obtaining a stable trace because of the low amplitude of the signal. How could you use the *output* signal from the preamplifier to help obtain a stable trace?

2. Refer to Test Bench 1 in the text showing waveforms 3 and 4. Notice that the oscilloscope is triggered from channel 2. Why is it easier to obtain a stable trace by triggering the oscilloscope on channel 2 rather than on channel 1?

19 Pulse Measurements

OBJECTIVES:

After performing this experiment, you will be able to:

1. Measure rise time, fall time, pulse repetition time, pulse width, and duty cycle for a pulse waveform.
2. Explain the limitations of instrumentation in making pulse measurements.
3. Compute the oscilloscope bandwidth necessary to make a rise time measurement with an accuracy of 3%.

READING:

Floyd, *Principles of Electric Circuits,* Sections 11–8 and 11–9

MATERIALS NEEDED:

One 1000 pF capacitor

Application Problem: One 2N3904 transistor, one 10 μF capacitor, resistors: 1 kΩ, 10 kΩ, 4.7 MΩ

SUMMARY OF THEORY:

A pulse is a signal that rises from one level to another, remains at the second level for some time, then returns to the original level. Definitions for pulses are illustrated in Figure 19–1. The time from one pulse to the next is the period, T. The reciprocal of the period is called the *pulse repetition frequency,* PRF. The time required for a pulse to rise from 10% to 90% of its maximum level is called the *rise time* and the time to return from 90% to 10% of the maximum level is called the *fall time.* Pulse width, abbreviated t_w, is measured at the 50% level, as illustrated. The duty cycle is the ratio of the pulse width to the period, usually expressed as a percentage

$$\text{Percent duty cycle} = \left(\frac{t_w}{T}\right) 100\%$$

Actual pulses differ from the idealized model shown in Figure 19–1(a). They may have *sag, overshoot,* or *undershoot* as illustrated in Figure 19–1(b). In addition, if cables are mismatched in the system, *ringing* may be observed. Ringing is the appearance of a short oscillatory transient that appears at the top and bottom of a pulse, as illustrated in Figure 19–1(c).

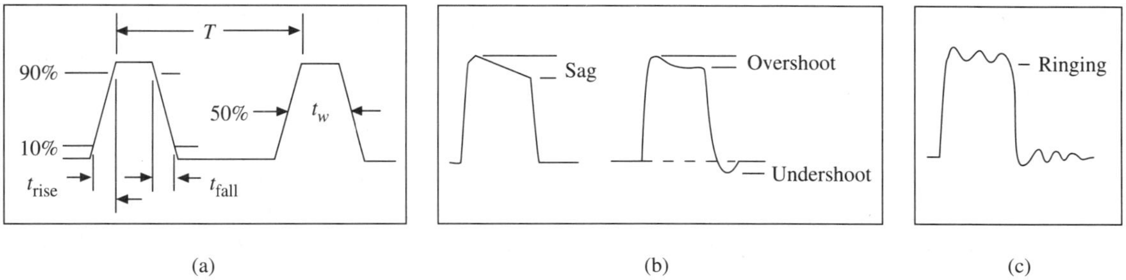

Figure 19–1

All measurements involve some error due to the limitations of the measurement instrument. Oscilloscopes can distort a pulse waveform including the distortions illustrated in Figure 19–1. In addition, if the rise time of the oscilloscope amplifier is too slow, rise time distortion may occur. *The oscilloscope rise time should be at least four times faster than the pulse to be measured if the observed rise time is to have less than 3% error.* If the oscilloscope rise time is only twice as fast as the measured rise time, the measurement error rises to over 12%! The oscilloscope rise time is related to the range of useful frequencies it can pass. This range of frequencies is called the *bandwidth.* To find the rise time of an oscilloscope when the bandwidth is known, the following approximate relationship is useful:

$$t_{(r)\text{ scope}} = \frac{0.35}{BW}$$

where $t_{(r)\text{scope}}$ is the oscilloscope rise time in microseconds and BW is the bandwidth in megahertz. For example, an oscilloscope with a 60 MHz bandwidth has a rise time of approximately 6 ns. Measurements of pulses with rise times faster than about 24 ns on this oscilloscope will have measurable error. A correction to the measured value can be applied to obtain the actual rise time of a pulse. The correction formula is

$$t_{(r)\text{ true}} = \sqrt{t_{(r)}^{2}{}_{\text{displayed}} - t_{(r)}^{2}{}_{\text{scope}}}$$

where $t_{(r)\text{ true}}$ is the actual rise time of the pulse, $t_{(r)\text{ displayed}}$ is the observed rise time, and $t_{(r)\text{ scope}}$ is the rise time of the oscilloscope. This formula can be applied to correct observed rise times by 10% or less.

Measurement of pulses normally should be done with the input signal coupled to the scope using dc coupling. This directly couples the signal to the oscilloscope and avoids causing pulse sag which can cause measurement error. Probe compensation should be checked before making pulse measurements. It is particularly important in rise time measurements to check probe compensation. This check is described in this experiment. It is also important in time and voltage measurements to be sure that the appropriate VARIABLE (VAR) controls (usually red) are in their calibrate position. Most oscilloscopes have a "detent" or notched position for these controls so the user is aware when they are calibrated.

PROCEDURE:

1. From the manufacturer's specifications, find the bandwidth of the oscilloscope you are using. Normally the bandwidth is specified with a 10X probe connected to the input. You should make oscilloscope measurements with the 10X probe connected to avoid bandwidth reduction. Use the specified bandwidth to compute the rise time of the oscilloscope as explained in the Summary of Theory. This will give you an idea of the limitations of the oscilloscope you are using to make accurate rise time measurements. Enter the bandwidth and rise time of the scope in Table 19–1.

2. Look on your oscilloscope for a probe compensation output. This output provides an internally generated square wave, usually at a frequency of 1.0 kHz. It is a good idea to check this signal when starting with an instrument to be sure that the probe is properly compensated. To compensate the probe, set the VOLTS/DIV control to view the square

wave over several divisions of the display. An adjustment screw on the probe is used to obtain a good square wave with a flat top. An improperly compensated oscilloscope will produce inaccurate measurements. If directed by your instructor, adjust the probe compensation.

3. Set the function generator for a square wave at a frequency of 100 kHz and an amplitude of 4.0 V. A square wave cannot be measured accurately with your meter—you will need to measure the voltage with the oscilloscope. Check the 0 V level on the oscilloscope and adjust the generator to go from 0 V to 4.0 V. Most function generators have a separate control to adjust the dc level of the signal.

4. Measure the parameters listed in Table 19–2 for the square wave from the function generator. Be sure the oscilloscope's SEC/DIV is in its calibrated position. If your oscilloscope has percent markers etched on the front graticule, you may want to *uncalibrate* the VOLTS/DIV when making rise and fall time measurements. Use the vertical POSITION control and VOLT/DIV vernier to position the waveform between 0% and 100% markers on the oscilloscope display. Then measure the time between the 10% and 90% markers.

5. To obtain practice measuring rise time, place a 1000 pF capacitor across the generator output. Measure the new rise and fall times. Record your results in Table 19–3.

6. If you have a separate pulse output from your function generator, measure the pulse characteristics listed in Table 19–4. To obtain good results with fast signals, the generator should be terminated in its characteristic impedance (typically 50 Ω). You will need to use the fastest sweep time available on your oscilloscope. Record your results in Table 19–4.

FOR FURTHER INVESTIGATION:

In many applications, it is important to measure time differences. One important technique for doing this is to use *delayed sweep* measurements. If your scope is equipped with delayed sweep, you can trigger from a signal and view a magnified portion of the signal at a later time. With dual time base oscilloscopes, delayed sweep offers increased timing accuracy. If you have a *calibrated* DELAY TIME POSITION dial, you can make differential delay time measurements between two different signals. Most delayed sweep oscilloscopes will have a HORIZONTAL MODE switch, which allows you to view the A sweep, the B sweep, or A intensified by B. The sweep speeds for A and B can be separately controlled, often by concentric rings on the SEC/DIV control. Consult the operator's manual for your oscilloscope to determine the exact procedure.* Then practice by measuring the pulse width at a frequency of 1 kHz from the pulse generator using delayed sweep. Summarize your procedure and results.

---

*An excellent source of information is: *The XYZs of Using an Oscilloscope,* Tektronix, Inc., PO Box 500, Beaverton, OR 97077.

APPLICATION PROBLEM:

The bandwidth of a circuit is the range of frequencies it can pass as illustrated in Figure 19–2. If the lower frequency is very low or dc, the bandwidth is approximately equal to the upper frequency at which the output has dropped to 70.7% of the midband frequencies. The bandwidth of such an amplifier can be found by maintaining a sinusoidal signal at a constant level on the input and increasing the frequency until the output has dropped to 70.7% of the midband level. This frequency can then be measured with an oscilloscope.

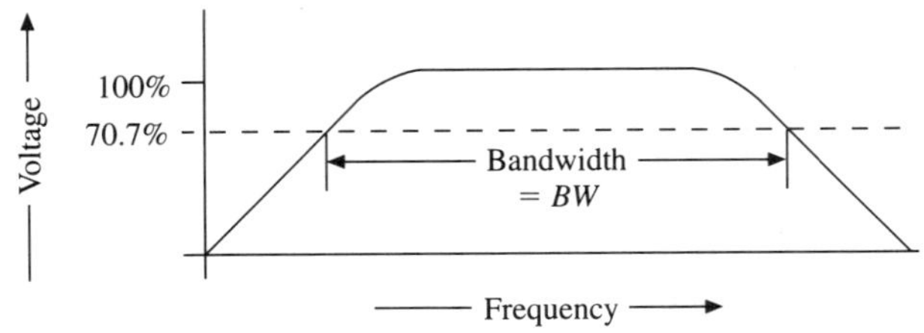

Figure 19–2

An alternative method for determining the bandwidth is to use a relatively fast rising square wave on the input and measure the rise time $t_{(r)}$ of the output signal. This method is good only for circuits with flat frequency responses, such as audio amplifiers. The bandwidth is given by

$$BW = \frac{0.35}{t_{(r)}}$$

Construct the transistor amplifier shown in Figure 19–3. Set the function generator for a 250 mV_{pp} sinusoidal wave at 100 kHz. It is a good idea to monitor both the input and output waveform to assure that the waveform into the circuit under test is undistorted. Measure the output voltage at the collector of the transistor. Then, while observing the output on an oscilloscope, raise the generator frequency until the output drops to 70.7%. You can use the 0% to 100% marks on the edge of the graticule for this. Measure this frequency and report it as the bandwidth of the circuit.

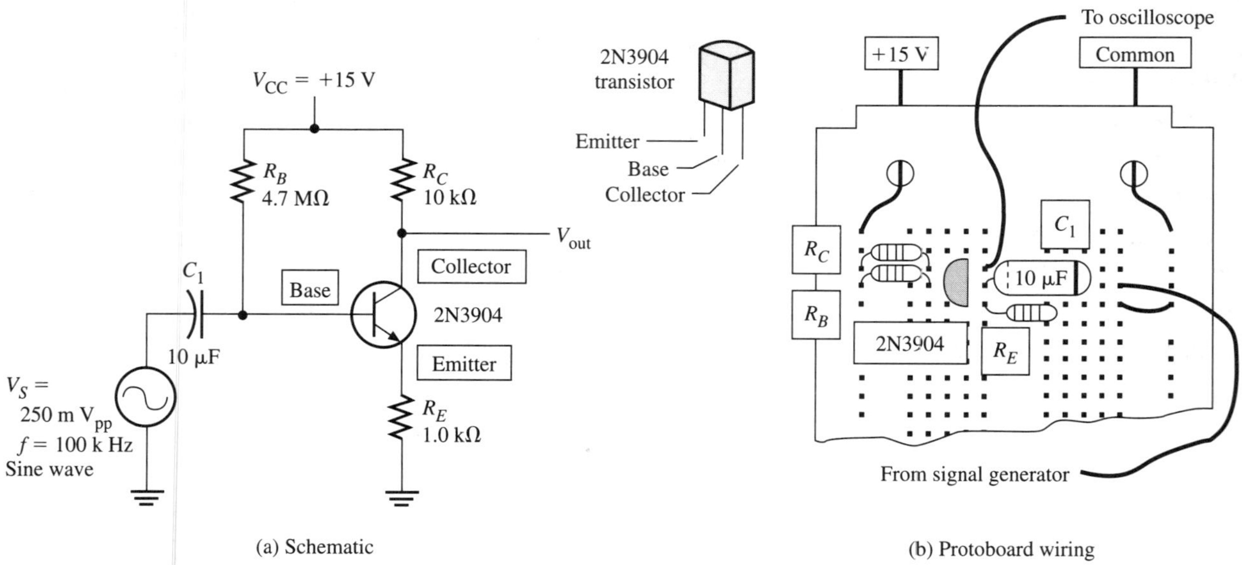

Figure 19–3

Change the input signal from a sine wave into a square wave at 100 kHz. Measure the rise time of the output square wave and use this measurement to determine the bandwidth. Compare the two methods in your report.

Report for Experiment 19

Name ______________
Date ______________
Class ______________

OBJECTIVES:

PROCEDURE:

DATA:

Table 19–1
Oscilloscope

BW =	
$t_{(r)}$ =	

Table 19–2
Function Generator
(square wave output)

Rise time, $t_{(r)}$	
Fall time, $t_{(f)}$	
Period, T	
Pulse width, t_w	
% duty cycle	

Table 19–3
Function Generator
(with 1000 pF capacitor across output)

Rise time, $t_{(r)}$	
Fall time, $t_{(f)}$	

Table 19–4
Function Generator
(pulse output)

Rise time, $t_{(r)}$	
Fall time, $t_{(f)}$	
Period, T	
Pulse width, t_w	
% duty cycle	

RESULTS AND CONCLUSION:

FURTHER INVESTIGATION RESULTS:

APPLICATION PROBLEM RESULTS:

EVALUATION AND REVIEW QUESTIONS:

1. Were any of the measurements limited by the bandwidth of the oscilloscope? If so, which ones?

2. If you need to measure a pulse with a predicted rise time of 10 ns, what bandwidth should the oscilloscope have to measure the time within 3%? (Hint: the rise time of the scope should be 4X better than the pulse rise time.)

3. The SEC/DIV control on many oscilloscopes has a X10 magnifier. When the magnifier is ON, the time scale must be divided by 10. Explain.

4. An oscilloscope presentation has the SEC/DIV control set to 2.0 ms/div and the X10 magnifier is OFF. Determine the rise time of the pulse shown in Figure 19–4.

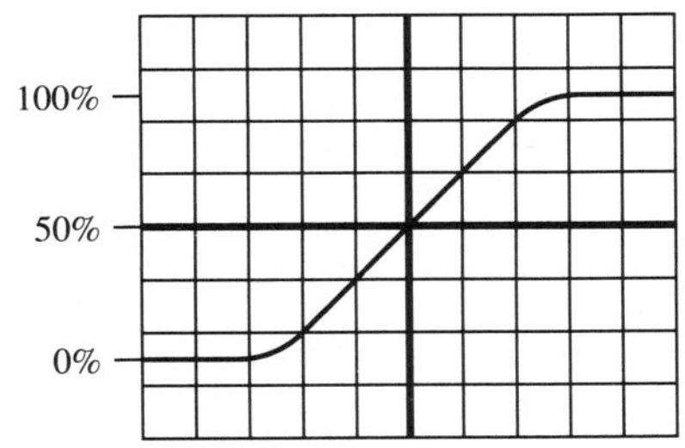

Figure 19–4

5. Repeat question 4 if the X10 magnifier had been ON.

6. Why are pulse measurements normally done using dc coupling for the signal?

20 Capacitors

OBJECTIVES:

After performing this experiment, you will be able to:

1. Compare total capacitance, charge, and voltage drop for capacitors connected in series and in parallel.
2. Test capacitors with an ohmmeter and a voltmeter as a basic charging test.

READING:

Floyd, *Principles of Electric Circuits,* Sections 13–1 through 13–5 and 13–8

MATERIALS NEEDED:

Two LEDs
Resistors: Two 1 kΩ
Capacitors: One of each: 100μF, 47 μF, 1.0 μF, 0.1 μF, 0.01 μF (35 WV or greater)
Application Problem: One additional 100 μF capacitor, one 100 kΩ resistor

SUMMARY OF THEORY:

A capacitor is formed whenever two conductors are separated by an insulating material. When a voltage exists between the conductors, there will be an electric charge between the conductors. The ability to store an electric charge is a fundamental property of capacitors and affects both dc and ac circuits. Capacitors are made with large flat conductors called *plates*. The plates are separated with an insulating material called a *dielectric*. The ability to store charge increases with larger plate size and closer separation.

When a voltage is connected across a capacitor, charge will flow in the external circuit until the voltage across the capacitor is equal to the applied voltage. The charge that flows is proportional to the size of the capacitor and the applied voltage. This is a fundamental concept for capacitors and is given by the equation:

$$Q = CV$$

where Q is the charge in coulombs, C is the capacitance in farads, and V is the applied voltage. An analogous situation is that of putting compressed air into a bottle. The quantity of air is directly proportional to the capacity of the bottle and the applied pressure.

Recall that current is defined as charge per time; that is,

$$I = \frac{Q}{t}$$

where I is the current in amperes, Q is the charge in coulombs, and t is the time in seconds. This equation can be rearranged as

$$Q = It$$

If we connect two capacitors in series with a voltage source, the same charging current flows through both capacitors. Since this current flows for the same amount of time, it can be seen that the total charge, Q_T, must be the same as the charge on each capacitor; that is,

$$Q_T = Q_1 = Q_2$$

Charging capacitors in series causes the same charge to be across each capacitor: however, as shown in Floyd's text, the total capacitance *decreases*. In a series circuit, the total capacitance is given by the formula

$$\frac{1}{C_T} = \frac{1}{C_1} + \frac{1}{C_2} + \ldots + \frac{1}{C_i}$$

Now consider capacitors in parallel. In a parallel circuit, the total current is equal to the sum of the currents in each branch as stated by Kirchhoff's current law. If this current flows for the same amount of time, the total charge leaving the voltage source will equal the sum of the charges which flow in each branch. Mathematically,

$$Q_T = Q_1 + Q_2 + \ldots + Q_i$$

Capacitors connected in parallel will raise the total capacitance because more charge can be stored at a given voltage. The equation for the total capacitance of parallel capacitors is:

$$C_T = C_1 + C_2 + \ldots + C_i$$

There are two quick tests that can verify that a capacitor, larger than about 0.01 μF, can be charged. Although the two tests are not comprehensive, they are useful in troubleshooting a faulty capacitor. The first test uses only an ohmmeter as a visual indication of charging to a small voltage (the internal voltage in the ohmmeter). An analog ohmmeter is best for this test. This test is done as follows:

1. Remove one end of the capacitor from the circuit and discharge it by placing a short across its terminals.
2. Set the ohmmeter on a high resistance scale and place the negative lead from an ohmmeter on the negative terminal of the capacitor. You must connect the ohmmeter with the proper polarity. *Do not assume the common lead from the ohmmeter is the negative side!*
3. Touch the other lead of the ohmmeter onto the remaining terminal of the capacitor. The meter should indicate very low resistance and then gradually increase resistance. If you put the meter in a higher range, the ohmmeter charges the capacitor slower and the capacitance "kick" will be emphasized. For small capacitors (under 0.01 μF), this change may not be seen. Large electrolytic capacitors require more time to charge, so use a lower range on your ohmmeter. Capacitors should never remain near zero resistance, as this indicates a short. An immediate high-resistance reading indicates an open for larger capacitors.

A capacitor that passes the ohmmeter test may fail when working voltage is applied. A voltmeter can be used to check a capacitor with voltage applied. The voltmeter is connected in *series* with the capacitor and a dc voltage as indicated in Figure 20–1. When voltage is first applied, the capacitor charges through the voltmeter's large series resistance. As it charges, voltage will appear across it, and the voltmeter indication will soon show a very small voltage. Large electrolytic capacitors may have leakage current that makes them appear bad, especially with a very high impedance voltmeter. In this case, use the test as a relative test, comparing the reading with a similar capacitor which you know is good.

The simple charging tests are satisfactory for determining if a gross failure has occurred. They do not indicate the value of the capacitor or if its value has changed. Value change is a common fault in capacitors, and there are other failures, such as high leakage current and dielectric absorption (the result of internal dipoles remaining in a polarized state even after the capacitor discharges). Some low cost DMMs include built-in capacitance meters. A more comprehensive test can be provided by an instrument such as a dynamic *component analyzer,* which measures the value as well as leakage current and dielectric absorption.

PROCEDURE:

1. Obtain 5 capacitors as listed in Table 20–1. Check each capacitor using the ohmmeter test described in the Summary of Theory. Record the results of the test on Table 20–1.

2. Test each capacitor using the voltmeter test. Because of slow charging, a large electrolytic capacitor may appear to fail this test. Check the voltage rating on the capacitor to be sure it is not exceeded. The working voltage is the maximum voltage that can safely be applied to the capacitor. Record your results in Table 20–1.

3. Connect the circuit shown in Figure 20–2. The switches can be made from wire. Leave both switches open. The light-emitting diodes (LEDs) and the capacitor are both polarized components—they must be connected in the correct direction in order to work properly.

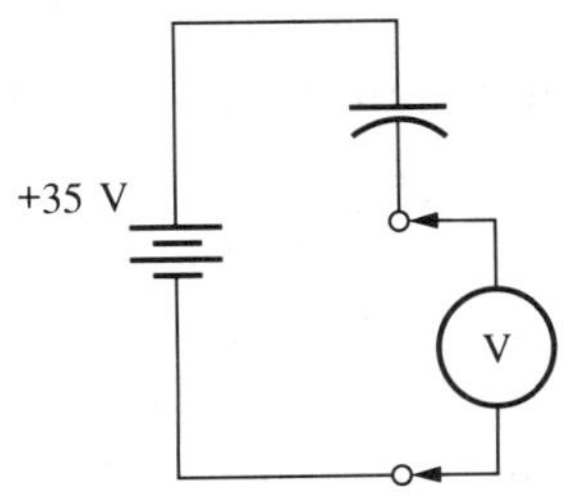

Figure 20–1

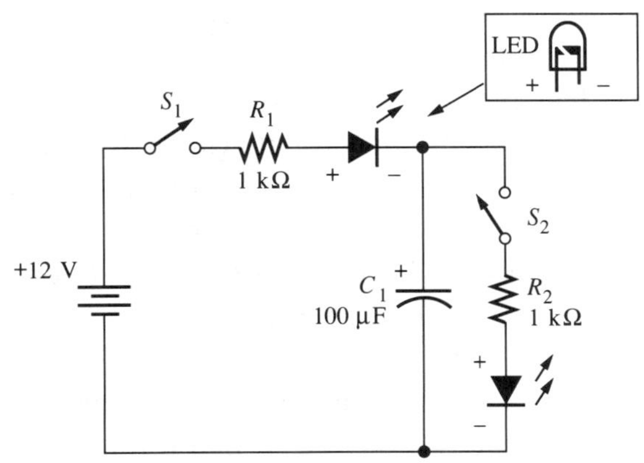

Figure 20–2

4. Close S_1 and observe the LEDs. Then open S_1 and then close S_2. Describe your observations in Table 20–2.

5. Now connect C_2 in series with C_1. Open both switches. Make certain the capacitors are fully discharged by shorting them with a piece of wire; then close S_1. Measure the voltage across each capacitor. Do this quickly to prevent the meter from causing the capacitors to discharge. Record the voltages in Table 20–2.

6. Using the measured voltages, compute the charge on each capacitor. Then open S_1 and close S_2. Record the computed charge and your observations in Table 20–2.

7. Change the capacitors from series to parallel. Open both switches. Ensure the capacitors are fully discharged. Then close S_1. Measure the voltage (quickly) across the parallel capacitors and enter the measured voltage in Table 20–2.

8. Using the measured voltage across the parallel capacitors, compute the charge on each one. Then open S_1 and close S_2. Record the computed charge and your observations in Table 20–2.

9. Replace the +12 V dc source with a signal generator. Set the signal generator to a square wave and set the amplitude to 12 V_{pp}. Set the frequency to 10 Hz. Close both switches. Notice the difference in the LED pulses. This demonstrates one of the principal applications of large capacitors—that of filtering. Record your observations.

FOR FURTHER INVESTIGATION:
Use the oscilloscope to measure the waveforms across the capacitors and the LEDs in step 9. Try speeding up the signal generator and observe the waveforms. Use the two-channel-difference measurement explained in Experiment 15 to see the waveform across the ungrounded LED. Draw and label the waveforms in your report.

APPLICATION PROBLEM:
A voltage multiplier is a circuit that uses diodes and capacitors to increase the peak value of a sine wave. Voltage multipliers can produce high voltages without requiring a high-voltage transformer. The circuit illustrated in Figure 20–3 is a full-wave voltage doubler. The circuit is drawn as a bridge with diodes in two arms and capacitors in two arms. The diodes allow current to flow in only one direction, charging the capacitors to near the peak voltage of the sine wave. Generally, voltage doublers are used with 60 Hz power line frequencies and with ordinary diodes, but in order to clarify the operation of this circuit, you can use the LEDs that were used in this experiment. (Note that the output voltage will be reduced.) Connect the circuit, setting the function generator to 20 V_{pp} sine wave at a frequency of 1 Hz. (If you cannot obtain a 20 V_{pp} signal, use the largest signal you can obtain from your generator.) Observe the operation of the circuit; then try speeding up the generator. Look at the waveform across the load resistor with your oscilloscope using the two-channel-difference method. What is the dc voltage across the load resistor? What happens to the output as the generator is speeded up? Try a smaller load resistor. Can you explain your observations?

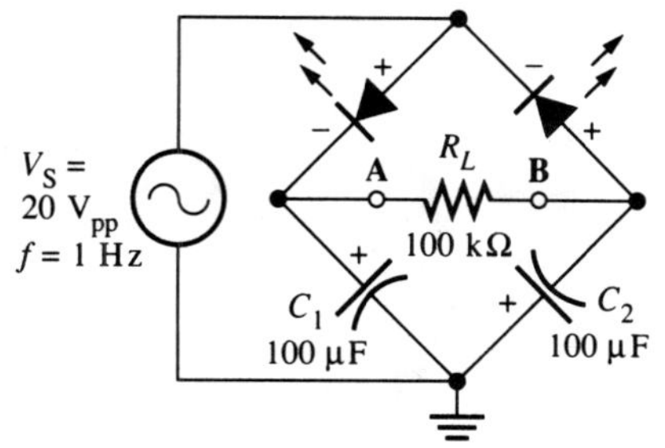

Figure 20–3

Report for Experiment 20

Name ______________
Date ______________
Class ______________

OBJECTIVES:

PROCEDURE:

DATA:

Table 20–1

Capacitor	Listed Value	Ohmmeter Test Pass/Fail	Voltmeter Test Pass/Fail
C_1	100 μF		
C_2	47 μF		
C_3	1.0 μF		
C_4	0.1 μF		
C_5	0.01 μF		

Table 20–2

Step	V_1	V_2	Q_1	Q_2	Observations
4					
5					
6					
7					
8					
9					

RESULTS AND CONCLUSION:

FURTHER INVESTIGATION RESULTS:

APPLICATION PROBLEM RESULTS:

EVALUATION AND REVIEW QUESTIONS:

1. Why did the LEDs flash for a shorter time in steps 5 and 6 than in step 4?

2. What would happen if you added more series capacitance in step 6?

3. (a) What is the total capacitance when a 1.0 μF capacitor is connected in parallel with a 2.0 μF capacitor?

   (b) If the capacitors are connected in series, what is the total capacitance?

   (c) In the preceding series connection, which capacitor has the greater voltage across it?

4. A 3 μF capacitor is charged to 100 V. If it is then connected in *parallel* with a 10 μF capacitor, what voltage will be across the capacitors? (Hint: total charge is conserved.)

5. Two capacitors, labeled A and B, are connected in series. They have the same charge but A has twice the voltage across it than B has. Which capacitor has the greatest capacitance? Prove your answer.

6. What are the limitations of using an ohmmeter or voltmeter to test a capacitor?

21 Capacitive Reactance

OBJECTIVES:
After performing this experiment, you will be able to:
1. Measure the capacitive reactance of a capacitor at a specified frequency.
2. Compare the reactance of capacitors connected in series and parallel.

READING:
Floyd, *Principles of Electric Circuits,* Sections 13–6, 13–7, and 13–9

MATERIALS NEEDED:
Capacitors:
One of each: 0.1 μF, 0.047 μF
Resistors:
One of each: 1.0 kΩ
For Further Investigation: One 1.0 μF capacitor, one 4.7 kΩ resistor, one 10.0 kΩ resistor
Application Problem: One 100 kΩ resistor, one capacitor (value to be determined by student)

SUMMARY OF THEORY:
If a resistor is connected across a sine-wave generator, a current flows that is *in phase* with the applied voltage. If, instead of a resistor, we connect a capacitor across the generator, the current is not in phase with the voltage. This is illustrated in Figure 21–1. Note that the current and voltage have exactly the same frequency, but the current is *leading* the voltage by ¼ cycle.

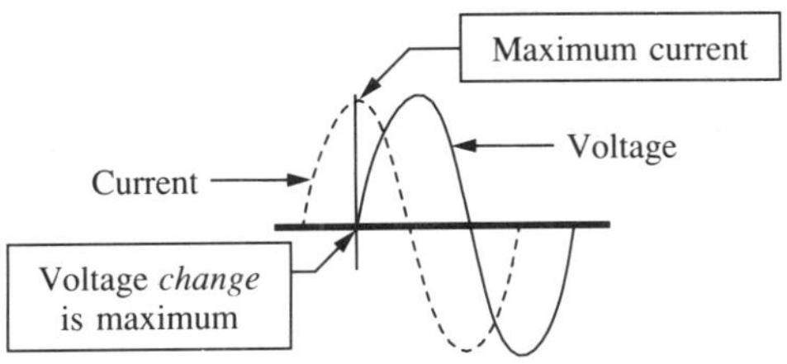

Figure 21–1 Current and voltage relationship in a capacitor.

Current flow in the capacitor is directly proportional to the capacitance and the rate of change of voltage. The largest current flows when the voltage *change* is a maximum. If the capacitance is increased or the frequency is increased, more current will flow. This is why a capacitor is sometimes thought of as a high-frequency short.

Reactance is the opposition to ac current flow and is measured in ohms, like resistance. Capacitive reactance is written with the symbol X_C. It can be defined as

$$X_C = \frac{1}{2\pi fC}$$

where f is the generator frequency in hertz and C is the capacitance in farads.

Ohm's law can be generalized to ac circuits. For a capacitor, we find the voltage across the capacitor using the current through the capacitor and the capacitive reactance. Ohm's law for the voltage across a capacitor is written

$$V_C = I_C X_C$$

PROCEDURE:

1. Obtain two capacitors with the values shown in Table 21–1. If you have a capacitance bridge available, measure their capacitance and record in Table 21–1; otherwise record the listed value of the capacitors. Measure and record the value of resistor R_1.

2. Set up the circuit shown in Figure 21–2. Set the function generator for a 1.0 kHz sine wave with a 1.0 V_{rms} output. Measure the rms voltage with your DMM while it is connected to the circuit. Check the frequency and voltage with the oscilloscope. *Note:* 1.0 V_{rms} = 2.828 V_{pp}.

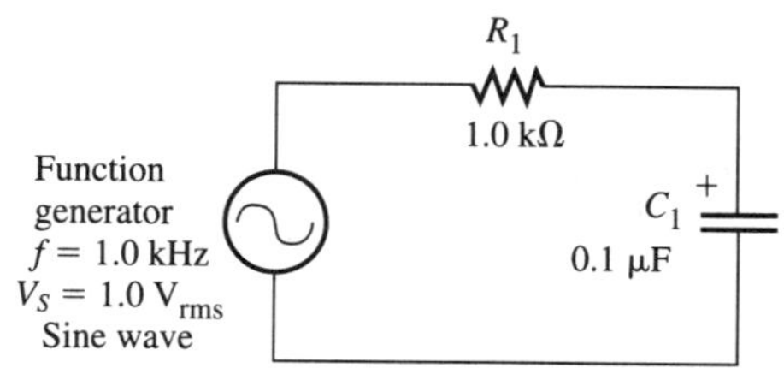

Figure 21–2

3. The circuit is a series circuit, so the current in the resistor and the capacitor are identical ($I_R = I_C = I$). You can find this current easily by applying Ohm's law to the resistor. Measure the voltage across the resistor, V_R, using the DMM. Record the measured voltage in Table 21–2 in the column labeled *Capacitor* C_1. Compute the current in the circuit by dividing the measured voltage by the resistance of R_1 and enter the value in Table 21–2.

4. Measure the rms voltage across the capacitor, V_C. Record this voltage in Table 21–2. Then use this voltage to compute the capacitive reactance using Ohm's law:

$$X_C = \frac{V_C}{I}$$

   Enter this value as the capacitive reactance in Table 21–2.

5. Using the capacitive reactance found in step 4, compute the capacitance using the equation

$$C = \frac{1}{2\pi f X_C}$$

Enter the computed capacitance in Table 21–2. This value should agree with the value marked on the capacitor and measured in step 1 within experimental tolerances.

6. Repeat steps 3, 4, and 5 using capacitor C_2. Enter the data in Table 21–2 in the column labeled *Capacitor* C_2.

7. Now connect C_1 in series with C_2. The equivalent capacitive reactance and capacitance can be found for the series connection by measuring across both capacitors as if they were one capacitor. Enter the data in Table 21–3 in the column labeled *Series Capacitors*. The following steps will guide you:
   (a) Check that the function generator is set to 1.0 V_{rms}. Find the current in the circuit by measuring the voltage across the resistor as before and dividing by the resistance. Enter the measured voltage and the current you found in Table 21–3.
   (b) Measure the voltage across *both* capacitors. Enter this voltage in Table 21–3.
   (c) Use Ohm's law to find the capacitive reactance of both capacitors. Use the voltage measured in step (b) and the current measured in step (a).
   (d) Compute the total capacitance by using the equation

$$C_T = \frac{1}{2\pi f X_{CT}}$$

8. Connect the capacitors in parallel and repeat step 7. Assume the parallel capacitors are one equivalent capacitor for the measurements. Enter the data in Table 21–3 in the column labeled *Parallel Capacitors*.

FOR FURTHER INVESTIGATION:
A capacitor can be used to couple an ac signal from one circuit to another. Typically, a dc voltage is also present, as shown in Figure 21–3. The ac and dc sources can be computed separately and algebraically combined as given by the superposition theorem. Use an oscilloscope to investigate this circuit by measuring the ac and dc voltages across each component. Remember to use the difference channel when making measurements across an ungrounded component. Show the dc and ac components of the voltage across R_1, R_2, and C_1. Change the frequency to 100 Hz and repeat the procedure. Summarize your results for both frequencies.

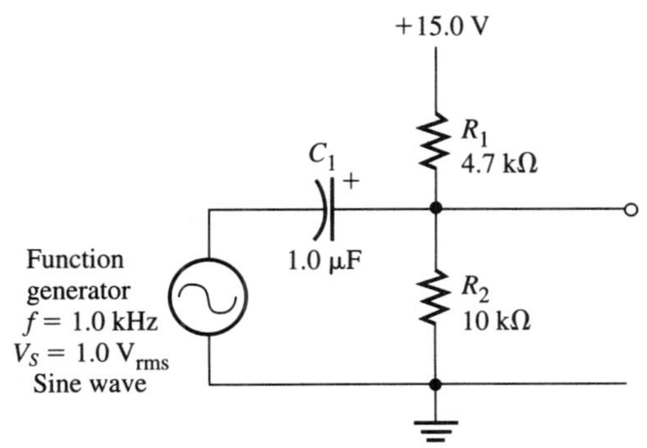

Figure 21–3

APPLICATION PROBLEM:

As illustrated in the Further Investigation, an application of capacitors is to couple an ac signal from one circuit to another while blocking any dc voltage. The capacitor is called a *coupling* capacitor. A coupling capacitor should look nearly like a short to the signal that is to be passed but appear open to the dc voltage. The basic coupling circuit is illustrated in Figure 21–4. R_{input} represents the input resistance of an amplifier and $C_{coupling}$ is the coupling capacitor.

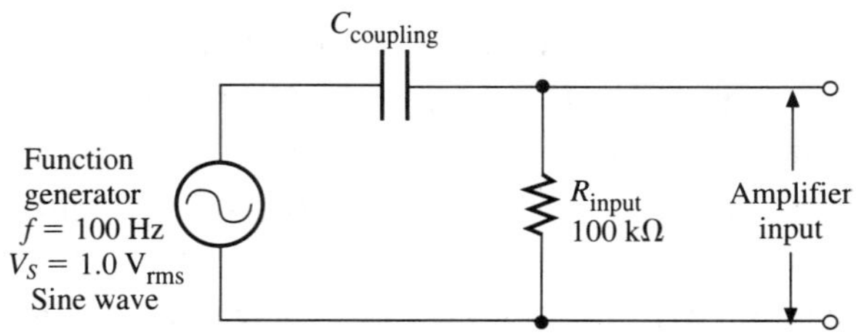

Figure 21–4

In this application, you need to find a capacitor that will allow a minimum of 90% of the generator signal to appear across R_{input} at a frequency of 100 Hz.

Compute the value of a capacitor that will meet this requirement. Construct your circuit and test it by measuring the generator voltage, the voltage drop across the capacitor, and the voltage drop across the resistor using a 100 Hz signal from the generator. Summarize your calculations and measurements in your report.

Report for Experiment 21

Name ____________
Date ____________
Class ____________

OBJECTIVES:

PROCEDURE:

DATA:

Table 21–1

Component	Listed Value	Measured Value
C_1	0.1 μF	
C_2	0.047 μF	
R_1	1.0 kΩ	

Table 21–2

	Capacitor C_1	Capacitor C_2
Voltage across R_1, (V_R)		
Total current, I		
Voltage across C, (V_C)		
Capacitive reactance, X_C		
Computed capacitance, C		

Table 21–3

Step		Series Capacitors	Parallel Capacitors
(a)	Voltage across R_1, V_R		
	Total Current, I		
(b)	Voltage across Capacitors, V_C		
(c)	Capacitive Reactance, X_C		
(d)	Computed Capacitance, C_T		

RESULTS AND CONCLUSION:

FURTHER INVESTIGATION RESULTS:

APPLICATION PROBLEM RESULTS:

EVALUATION AND REVIEW QUESTIONS:

1. Compare the capacitive reactance of the series capacitors with the capacitive reactance of the parallel capacitors. Use your data from Table 21–3.

2. Compare the total capacitance of the series capacitors with the total capacitance of the parallel capacitors.

3. If someone had mistakenly used too small a capacitor in a circuit, what would happen to the capacitive reactance?

4. Summarize the method used in this experiment to find the value of an unknown capacitor.

5. Assume the function generator is set to a higher (but known) frequency in this experiment. Will this affect the computed capacitance in this experiment? Explain:

6. Compute the capacitive reactance for a 800 pF capacitor at a frequency of 250 kHz.

TECH TIP REVIEW:
Each of the following questions relate to the TECH TIP circuit shown in Figure 13–54 of the Floyd text.

1. Assume that the input to the amplifier has no loading effect on the voltage divider. With this assumption, the voltage divider bias circuit can be replaced with a single equivalent resistance to ground. Determine the *frequency* at which the reactance of the coupling capacitor is equal to the resistance of the equivalent resistance of the voltage divider network. This frequency is called the cutoff frequency.

2. At the cutoff frequency determined in step 1, compare the amplitude of the signal at the input to the amplifier to the signal amplitude at the source.

3. What will happen to the signal amplitude at the input of the amplifier if the frequency of the input signal is lower than the cutoff frequency? Explain:

22 Inductors

OBJECTIVES:

After performing this experiment, you will be able to:

1. Describe the effect of Lenz's law in a circuit.
2. Measure the time constant of an *LR* circuit and test the effect of series and parallel inductances on the time constant.

READING:

Floyd, *Principles of Electric Circuits,* Sections 14–1 through 14–5

MATERIALS NEEDED:

Two 7 H inductors (approximate value) (Triad C-8X or equivalent) (second inductor may be shared with another experimenter)
One neon bulb (NE-2 or equivalent)
One 33 kΩ resistor
For Further Investigation: One unknown inductor
Application Problem: One 100 μF capacitor, one 1N4001 diode

SUMMARY OF THEORY:

When a current flows through a coil of wire, a magnetic field is created in the space surrounding the wire. This electromagnetic field accompanies any moving electric charge and is proportional to the magnitude of the current. If the current in the coil changes, the electromagnetic field causes a voltage to be induced across the coil which opposes the change. This property, which causes a voltage to oppose a change in current, is called *inductance*.

Inductance is the electrical equivalent of inertia in a mechanical system. It opposes a change in *current* in a manner similar to how capacitance opposed a change in *voltage*. This property of inductance is described by Lenz's law. According to Lenz's law, an inductor develops a voltage across it which counters the effect of a *change* in current in the circuit. Inductance is measured in *henries*. One henry is defined as the quantity of inductance present when one volt is generated as a result of a current changing at the rate of one ampere per second. Coils made to provide a specific amount of inductance are called *inductors*.

When inductors are connected in series, the total inductance is the sum of the individual inductors. This is similar to resistors connected in series. Likewise, the formula for parallel inductors is similar to the formula for parallel resistors. Unlike resistors, an additional effect can appear in inductive circuits. This effect is called *mutual inductance* and is caused by interaction of the magnetic fields. The total inductance can be either increased or decreased due to mutual inductance.

Inductive circuits have a time constant associated with them, just as capacitive circuits do, except the rising exponential curve is a picture of the *current* in the circuit rather than the *voltage,* as in the case of the capacitive circuit. Unlike the capacitive circuit, if the resistance is greater, the time constant is shorter. The time constant is found from the equation:

$$\tau = \frac{L}{R}$$

where τ = time constant in seconds,
L = inductance in henries and
R = resistance in ohms.

PROCEDURE:

1. In this step, you can observe the effect of Lenz's law. Connect the circuit shown in Figure 22–1 with a neon bulb in parallel with a large inductor. Neon bulbs contain two insulated electrodes in a glass envelope containing neon gas. The gas will not conduct unless the voltage reaches approximately 70 V. When the gas conducts, the bulb will glow. When the switch is closed, dc current in the inductor is determined by the inductor's winding resistance. Close and open S_1 several times and describe your observations in the report.

2. Find out if the neon bulb will fire if the voltage is lowered. How low can you reduce the voltage source and still observe the bulb? Record your observations in the report.

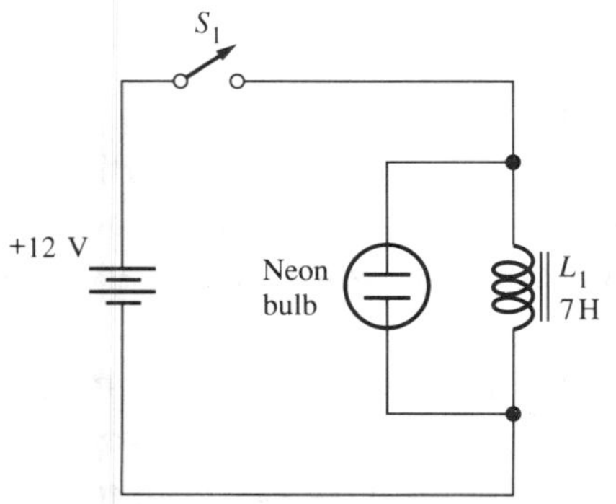

Figure 22–1

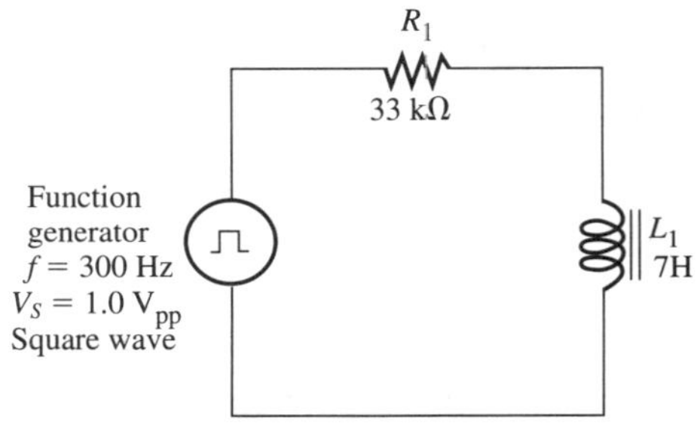

Figure 22–2

3. Connect the circuit shown in Figure 22–2. This circuit will be used to view the waveforms from a square-wave generator. Set the generator, V_S, for a 1.0 V_{pp} square wave at a frequency of 300 Hz. This frequency is chosen to allow sufficient time to see the effects of the time constant. View the generator voltage on CH1 of a two-channel oscilloscope and the inductor waveform on CH2. If both channels are calibrated and have the VOLTS/DIV controls set to the same setting, you will be able to see the voltage across the resistor using the difference channel. Set the oscilloscope SEC/DIV control to 0.5 ms/div. Sketch the waveforms you see on Plot 22–1.

4. Compute the time constant for the circuit. Enter the computed value in Table 22–1. Now measure the time constant by viewing the waveform across the resistor. The resistor voltage has the same shape as the current in the circuit, so you can measure the time constant by finding the time required for the resistor voltage to change from 0 to

63% of its final value.* Stretch the waveform across the oscilloscope screen to make an accurate time measurement. Enter the measured time constant in Table 22–1.

5. When inductors are connected in series, the total inductance increases. When they are connected in parallel, the total inductance decreases. To see the effect of parallel inductors, connect *one* end of a second 7 H inductor to the first inductor. Then, while observing the waveform across the first inductor, complete the parallel connection of the inductors. You should observe the waveform change as you alternately add or remove the parallel inductor.

   You can see the effect of series inductors by placing the two inductors in series. While observing the waveform across *both* inductors, short out one of the inductors with a jumper wire. Note what happens to the voltage waveforms across the resistor and the inductors. Describe your observations in the space provided in the report.

FOR FURTHER INVESTIGATION:

Suggest a method in which you could use a square-wave from a function generator and a known resistor to determine the inductance of an unknown inductor. Then obtain an unknown inductor from your instructor and measure its inductance. Report on your method, your result, and how your result compares to the accepted value for the inductor.

APPLICATION PROBLEM:

A switching power supply is a very efficient way to produce regulated dc. It uses transistor switches to rapidly "chop" unregulated dc into pulses. The transistors are controlled by circuits that sense the output voltage and vary the duty cycle of the pulses. This action causes the output voltage to be controlled by varying the on time of the pulses. The pulse train is smoothed by inductive and capacitor filters before the load.

The circuit shown in Figure 22–3 represents the output portion of a switching regulator. The pulse generator represents the output from the switching transistors. The inductor is normally a "pot-core" nonsaturable inductor. The inductor in this experiment is much larger than is usually needed; however, it will serve to illustrate the filtering action. When the pulse generator is ON, current flows through the inductor in the direction indicated in Figure 22–3(a).

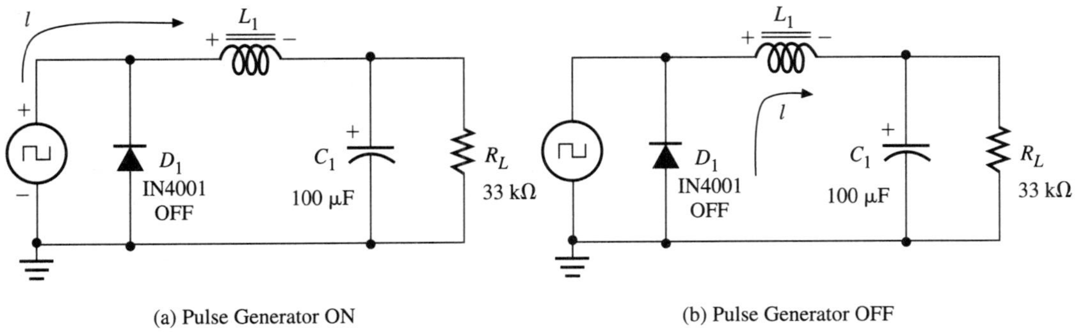

Figure 22–3

---

*Alternatively, you can measure the rise time and convert the reading to time constant. The relation between rise time and time constant is $\tau = \frac{t_r}{2.20}$.

Note that the diode is OFF. When the pulse generator is OFF, a voltage is induced across the inductor as shown in Figure 22–3(b). This voltage tends to keep the current flowing in the same direction. The diode is ON and provides a path for this current from ground.

Set up the circuit using a 5 V square wave from your function generator. Observe the waveform on your oscilloscope from the generator. At the same time, look at the voltage across the output. How much ripple is present? What does the waveform across the inductor look like? What happens to the ripple if the load resistor is smaller? Investigate these points and summarize your findings in your report.

Report for Experiment 22

Name ______________
Date ______________
Class ______________

OBJECTIVES:

PROCEDURE:

DATA:

Observations from Step 1:

Observations from Step 2:

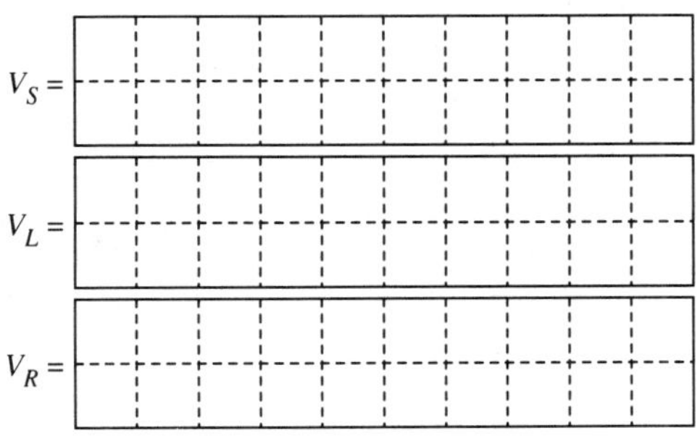

Plot 22–1

Table 22–1

	Computed Time	Measured Time
Time constant, τ		

Observations from Step 5:

RESULTS AND CONCLUSION:

FURTHER INVESTIGATION RESULTS:

APPLICATION PROBLEM RESULTS:

EVALUATION AND REVIEW QUESTIONS:

1. The ionizing voltage for a neon bulb is approximately 70 V. Explain how a 12 V source was able to cause the neon bulb to conduct.

2. When a circuit containing an inductor is opened suddenly, an arc may occur across the switch. How does Lenz's law explain this?

3. (a) What is the total inductance when two 100 mH inductors are connected in series?

   (b) In parallel?

4. What would happen to the time constant in Figure 22–2 if a 3.3 kΩ resistor were used instead of the 33 kΩ resistor?

5. What effect does an increase in the frequency of the square wave generator have on the waveforms observed in Figure 22–2?

6. State a rule for determining the polarity of the voltage induced across the inductor.

23 Inductive Reactance

OBJECTIVES:

After performing this experiment, you will be able to:

1. Measure the inductive reactance of an inductor at a specified frequency.
2. Compare the reactance of inductors connected in series and parallel.

READING:

Floyd, *Principles of Electric Circuits,* Sections 14–6 through 14–9

MATERIALS NEEDED:

Two 100 mH inductors
One 1.0 kΩ resistor
For Further Investigation: One 12.6 V center-tapped transformer
Application Problem: One 0.1 μF capacitor, two decade resistance boxes

SUMMARY OF THEORY:

When a sine wave is applied to an inductor, a voltage is induced across the inductor as given by Lenz's law. At a point where the *change* in current is a maximum, the largest induced voltage appears across the inductor. This is illustrated in Figure 23–1. Notice that when the current is not changing, the induced voltage is zero. For this reason, the voltage that appears across an inductor leads the current in the inductor by ¼ cycle.

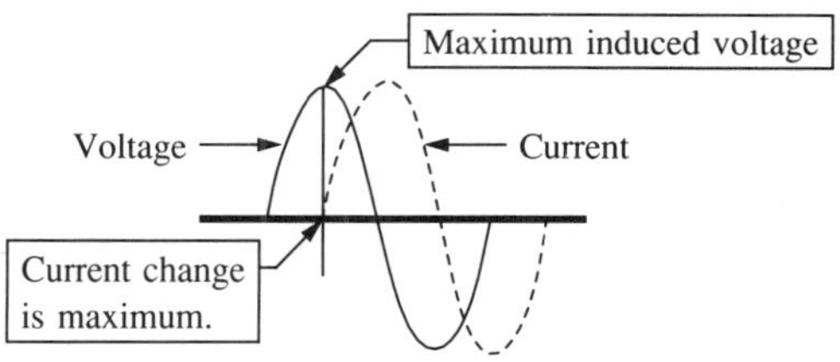

Figure 23–1 Current and voltage relationship in an inductor.

If we *raise* the frequency of the sine wave, the rate of change of current is increased and the value of the opposing voltage is increased. This results in a net decrease in the amount of current that flows. Thus the inductive reactance is increased by an increase in frequency. The inductive reactance is given by the equation

$$X_L = 2\pi f L$$

A linear relationship exists between the inductance and the reactance at a given frequency. Recall that in series, the total inductance is the sum of individual inductors (ignoring mutual inductance). The reactance of series inductors is, therefore, also the sum of the individual reactances.

Likewise, in parallel, the reciprocal formula which applies to parallel resistors can be applied to both the inductance and the inductive reactance.

Ohm's law may also be applied to inductive circuits. The reactance of an inductor can be found by dividing the voltage across the inductor by the current through the inductor; that is,

$$X_L = \frac{V_L}{I_L}$$

In this experiment, you will gain further practice making measurements with the oscilloscope. To simplify making oscilloscope measurements, the voltages and currents are all specified as peak-to-peak values. As long as you are consistent and record *all* measurements as peak-to-peak values, the reactance found by Ohm's law will be correct.

PROCEDURE:

1. Measure the inductance of each of two 100 mH inductors and record their measured values in Table 23–1. Measure and record the actual value of a 1.0 kΩ resistor. Use the listed values if you cannot measure the inductors.

2. Connect the circuit shown in Figure 23–2. Using the oscilloscope, set the generator for a 5.0 kHz sine wave at 1.0 V_{pp}. Check the frequency with the oscilloscope. Record all current and voltage readings using peak-to-peak values in this experiment.

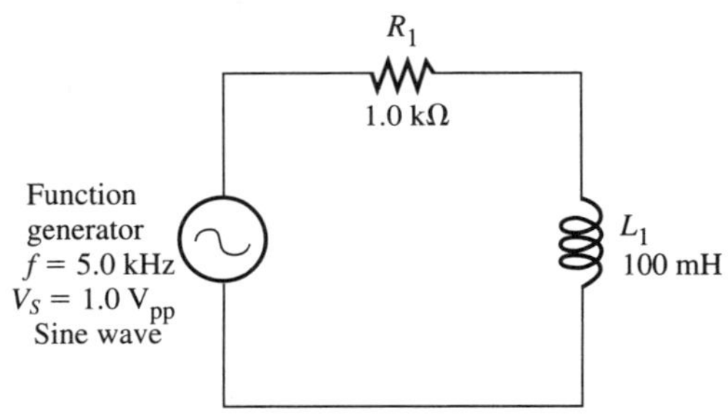

Figure 23–2

3. **The circuit is a series circuit, so the current in the resistor is the identical current that flows through the inductor ($I_R = I_L = I$). First, find the peak-to-peak voltage across the resistor using the two-channel-difference technique described in Experiment 18. Then apply Ohm's law to the resistor to find the peak-to-peak current in the circuit. Record the measured voltage and the computed current in Table 23–2 in the column labeled Inductor L_1.**

4. Measure the voltage across the inductor with the oscilloscope. Then find the inductive reactance by Ohm's law. Enter the values in Table 23–2.

5. Now compute the inductance from the equation

$$L = \frac{X_L}{2\pi f}$$

Enter the computed inductance in Table 23–2.

6. Replace L_1 with L_2 and repeat steps 3, 4, and 5. Enter the data in Table 23–2 in the column labeled Inductor L_2.

7. Place L_2 in series with L_1. Then find the inductive reactance for the series combination of the inductors as if they were one inductor. Enter the data in Table 23–3 in the column labeled Series Inductors. The following steps will guide you:
   (a) Check with the oscilloscope that the generator is set to 1.0 V_{pp}. Find the peak-to-peak current in the circuit by measuring the voltage across the resistor as before and dividing by the resistance.
   (b) Measure the peak-to-peak voltage across *both* inductors.
   (c) Use Ohm's law to find the inductive reactance of both inductors. Use the voltage measured in step (b) and the current found in step (a).
   (d) Compute the total inductance by using the equation

$$L = \frac{X_L}{2\pi f}$$

8. Connect the inductors in parallel and repeat step 7. Assume the parallel inductors are one equivalent inductor for the measurements. Enter the data in Table 23–3 in the column labeled Parallel Inductors.

FOR FURTHER INVESTIGATION:
A transformer consists of two or more coils wound on a common iron core. Frequently, one or more windings have a *center-tap,* which splits a winding into two equal inductors. Because the windings are on the same core, mutual inductance exists between the windings. Obtain a small power transformer that has a low-voltage center-tapped secondary winding. Determine the inductance of each half of the winding using the method in this experiment. Then investigate what happens if the windings are connected in series. Keep the output of the signal generator constant for the measurements. Summarize your results.

APPLICATION PROBLEM:
A Maxwell bridge is commonly used to measure inductors that do not have a very high Q. It employs a fixed capacitor and two resistors as standards. Capacitors are closer to being ideal circuit elements than are inductors because they have very little resistive losses. The circuit for a Maxwell bridge is shown in Figure 23–3. Construct the bridge using two decade resistance boxes for R_1 and R_2 and a measured capacitor of 0.1 μF for C_1. Set the frequency for 1.00 kHz and check it with the oscilloscope. The 100 mH inductor from this experiment (or any unknown from about 1 mH to 100 mH) can be used for the unknown inductor. R_3 is a fixed 1.0 kΩ resistor. Measure the output voltage between terminals **A** and **B** with your DMM. Adjust the decade boxes for the minimum voltage observed on the DMM.

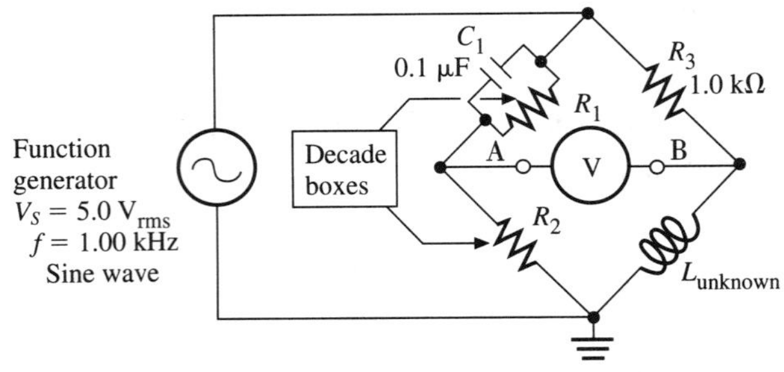

Figure 23–3

This circuit is an example of a balanced ac bridge. In any ac bridge, the bridge is balanced when the product of the impedance of the diagonal elements is equal. The equations for the Maxwell bridge, given without proof, are

$$L_{unknown} = R_2R_3C_1$$
$$Q = 2\pi fC_1R_1$$

Measure the unknown inductor with your Maxwell bridge. Compare your measurement with a laboratory bridge. What measurement errors account for the differences in the two measurements?

Report for Experiment 23

Name ______________
Date ______________
Class ______________

OBJECTIVES:

PROCEDURE:

DATA:

Table 23–1

Component	Listed Value	Measured Value
L_1	100 mH	
L_2	100 mH	
R_1	1.0 kΩ	

Table 23–2

	Inductor L_1	Inductor L_2
Voltage across R_1, V_R		
Total current, I		
Voltage across L, V_L		
Inductive Reactance, X_L		
Computed Inductance, L		

Table 23–3

Step		Series Inductors	Parallel Inductors
a	Voltage across R_1, V_R		
	Total current, I		
b	Voltage across inductors, V_L		
c	Inductive reactance, X_L		
d	Computed inductance, L		

RESULTS AND CONCLUSION:

FURTHER INVESTIGATION RESULTS:

APPLICATION PROBLEM RESULTS:

EVALUATION AND REVIEW QUESTIONS:

1. (a) Using the data in Table 23–2, compute the sum of the inductive reactances of the two inductors:

$$X_{L1} + X_{L2} =$$

(b) Using the data in Table 23–2, compute the product-over-sum of the inductive reactances of the two inductors:

$$\frac{(X_{L1})(X_{L2})}{X_{L1} + X_{L2}} =$$

(c) Compare the results from (a) and (b) with the reactances for the series and parallel connections listed in Table 23–3. What conclusion can you draw from these data?

2. Repeat question 1 using the data for the inductance, L. Compare the inductance of series and parallel inductors.

3. What effect would an error in the frequency of the generator have on the data for this experiment?

4. Explain how you could apply the method used in this experiment to find the inductance of an unknown inductor?

5. Compute the inductive reactance of a 50 μH inductor at a frequency of 500 kHz.

6. In this experiment, you were directed to record all voltages and currents as peak-to-peak values. How would recording the results as rms values have affected the inductance and the inductive reactance that you found?

TECH TIP REVIEW:
Each of the following questions relate to the TECH TIP circuit shown in Figure 14–41 of the Floyd text.

1. Explain how you would set up the oscilloscope to view the waveform across the inductor:

2. Draw the waveform you would expect to see for the voltage across the inductor:

24 Transformers

OBJECTIVES:

After performing this experiment, you will be able to:

1. Determine the turns ratio for a transformer.
2. Show the phase relationships between the primary and secondary of a center-tapped transformer.
3. Compute the turns ratio required for matching a signal generator to a speaker.
4. Demonstrate how an impedance matching transformer can increase the power transferred to a load.

READING:

Floyd, *Principles of Electric Circuits,* Sections 15–1 through 15–12

Materials Needed:

One 12.6 V center-tapped transformer (Triad F-70X or equivalent)
One small impedance matching transformer (approximately 600 Ω to 8 Ω)
One small speaker (4 or 8 Ω)
For Further Investigation: One 100 Ω resistor
Application Problem: IF transformer (J.W. Miller 8812 or equivalent)

SUMMARY OF THEORY:

A transformer consists of two (or more) closely coupled coils which share a common magnetic field. When an ac voltage is applied to the first coil, called the *primary,* a voltage is induced in the second coil, called the *secondary.* The voltage that appears across the secondary is proportional to the transformer turns ratio. The turns ratio is found by dividing the number of turns in the secondary winding by the number of turns in the primary winding. The turns ratio, n, is directly proportional to the primary and secondary voltages; that is,

$$n = \frac{N_S}{N_P} = \frac{V_S}{V_P}$$

For most work, we can assume that a transformer has no internal power dissipation and that all the magnetic flux lines in the primary also cut through the secondary—that is, we can assume the transformer is *ideal.* The ideal transformer delivers to the load 100% of the applied power. Actual transformers have losses due to magnetizing current, eddy currents, coil resistance, and so forth. In typical power applications, transformers are used to change the ac line voltage from one voltage to another or to isolate ac grounds. For the ideal transformer, the secondary voltage is found by multiplying the turns ratio by the applied primary voltage; that is,

$$V_S = nV_P$$

Since the ideal transformer has no internal losses, we can equate the power delivered to the primary to the power delivered by the secondary. Since $P = IV$, we can write

$$\text{Power} = I_P V_P = I_S V_S$$

This equation shows that if the transformer causes the secondary voltage to be higher than the primary voltage, the secondary current must be less than the primary current. Also, if the transformer has no load, then no primary or secondary current will flow in the ideal transformer.

In addition to their ability to change voltages and isolate grounds, transformers are useful to change the resistance (or *impedance*) of a load as viewed from the primary side. (Impedance is a more generalized word meaning opposition to ac current.) The load resistance appears to increase by the turns ratio squared (n^2) when viewed from the primary side. Transformers used to change impedance are designed differently than power transformers. They need to transform voltages over a band of frequencies with low distortion. Special transformers called *audio,* or *wideband,* transformers are designed for this. To find the correct turns ratio needed to match a load impedance to a source impedance, use the following equation:

$$n = \sqrt{\frac{R_{\text{LOAD}}}{R_{\text{SOURCE}}}}$$

This equation is based on the *ideal* transformer. Real transformers are *not* ideal. In this experiment, you will examine both a power transformer and an impedance matching transformer and calculate parameters for each.

PROCEDURE:

1. Obtain a low-voltage power transformer with a center-tapped secondary (12.6 V secondary). Using an ohmmeter, measure the primary and secondary resistance. Record the values in Table 24–1.

2. Compute the turns ratio based on the normal line voltage (V_P) of 115 V and the specified secondary voltage of 12.6 V. Record this as the computed turns ratio, *n,* in Table 24–1.

3. For safety, we will use a function generator set to 60 Hz in place of ac line voltages. Connect the circuit illustrated in Figure 24–1. Power transformers are designed to operate at a specific frequency (generally 60 Hz). Set the function generator to a 60 Hz sine wave at 5.0 V_{rms} on the primary. Measure the secondary voltage. From the measured voltages, compute the turns ratio for the transformer. Enter this value as the measured turns ratio in Table 24–1.

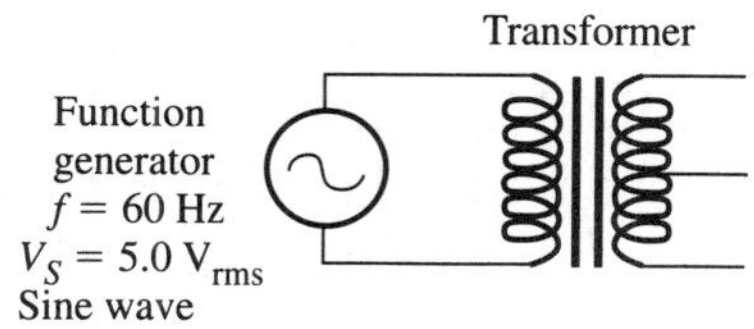

Figure 24–1

4. Compute the percent difference between the computed and measured turns ratio and enter the result in Table 24–1. The percent difference is found from the equation

$$\% \text{ diff} = \frac{n(\text{meas.}) - n(\text{comp.})}{n(\text{comp.})} \times 100$$

5. Connect a two-channel oscilloscope to the secondary as illustrated in Figure 24–2(a). Trigger the oscilloscope from channel 1. Do not use composite triggering. Compare the phase of the primary side viewed on channel 1 with the phase of the secondary side viewed on channel 2. Then reverse the leads on the secondary side. Describe your observations in your report.

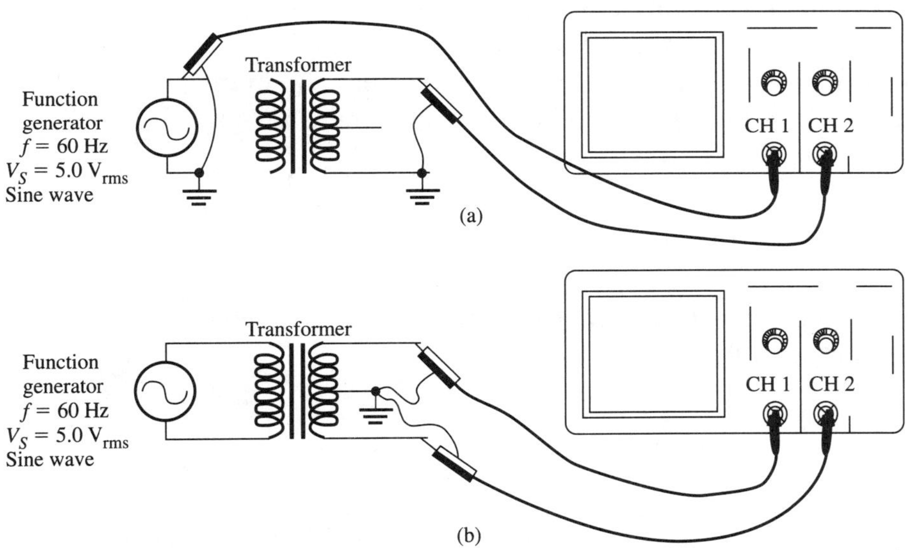

Figure 24–2

6. Connect the oscilloscope ground to the center-tap of the transformer and view the signals on each side at the center-tap at the same time, as illustrated in Figure 24–2(b). Describe the waveforms in your report.

7. In this step, a transformer will be used to match a source impedance to a load impedance. A small speaker represents a low impedance (typically 4 or 8 Ω), whereas a function generator is typically 600 Ω of driving impedance.* An impedance matching transformer can make the load appear to have the same impedance as the source. This allows maximum power to be

---

*The impedance of your generator, R_{GEN}, can be measured by the method described in Experiment 12, Further Investigation.

transferred to the load. Connect a small speaker directly to your function generator and set the frequency to approximately 2 kHz. Note the volume of the sound from the speaker. Measure the voltage across the speaker and record the measured voltage in Table 24–2.

8. Using the specified driving impedance of the function generator and the specified speaker impedance, compute the turns ratio required to match the speaker with your generator. Connect a small impedance matching transformer into the circuit. It is not necessary to obtain the precise turns ratio that you computed in order to hear the improvement in the power from the speaker. You can find the correct leads for the primary and secondary of the impedance-matching transformer using an ohmmeter. Since the required transformer is a step-down type, the primary resistance will be higher than the secondary resistance. Often, the primary winding will have a center-tap for push-pull amplifiers. Again measure the voltage across the speaker and record it in Table 24–2.

FOR FURTHER INVESTIGATION:

The ideal transformer model neglects a small current that flows in the primary independent of secondary load current. This current, called the *magnetizing* current, is required to produce the magnetic flux and is added to the current that is present due to the load. The magnetizing current appears to be flowing through an equivalent inductor parallel to the ideal transformer. Investigate this current by connecting the circuit shown in Figure 24–3 using the impedance-matching transformer. Calculate the magnetizing current, I_M, in the primary by measuring the voltage across a series resistor with no load and applying Ohm's law:

$$I_M = \frac{V_R}{R}$$

Find out if the magnetizing current changes as frequency is changed. Be sure to keep the generator at a constant 5.0 V_{rms}.

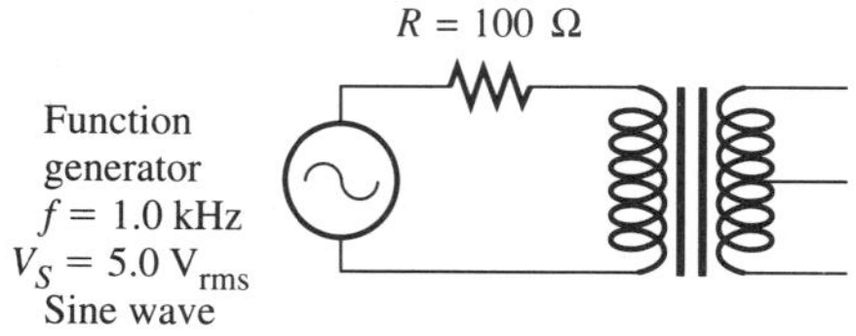

Figure 24–3

APPLICATION PROBLEM:

At radio frequencies, special transformers are widely used to couple a signal from one circuit to another. These transformers can be either *narrow-band,* meaning that they pass only a selected band of frequencies, or *broadband,* meaning that they pass a very wide range of frequencies. One type of narrow-band transformer used in receivers is called an IF (for *intermediate frequency*) transformer. One standard frequency for receiver IF transformers is 455 kHz.

Frequently, the primary of the IF transformer is in the collector circuit of a transistor amplifier as shown in Figure 24–4. The transistor amplifier couples its signal to the next stage through the IF transformer. The advantage of this technique is that the transistor amplifier is more efficient (as there is no load resistance to dissipate heat) and the transformer facilitates matching the impedance. At high frequencies, the size of the transformer is relatively small. Furthermore, the dc current required to operate the transistor is isolated from the next stage (or load) by the transformer.

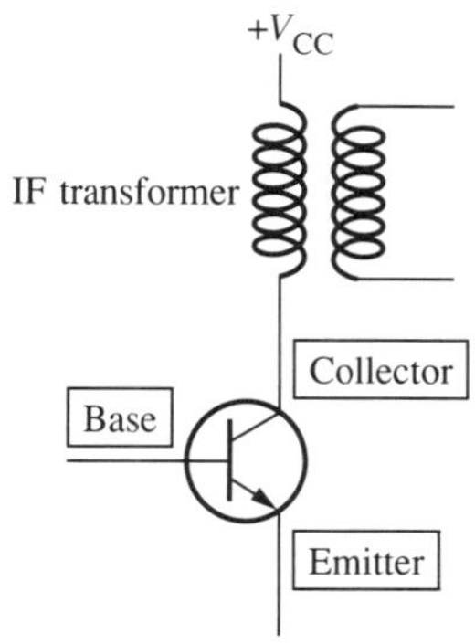

Figure 24–4

Assume you need an IF transformer for 455 kHz with 20 kΩ of input (primary) impedance and 5 kΩ of output (secondary) impedance. Determine the inductance required for the primary and the secondary coils and the turns ratio of the required transformer.

If you have an IF transformer available, measure the inductance of the primary and secondary windings, and determine the turns ratio. Summarize your calculations and measurements in your report.

Report for Experiment 24

Name ______________
Date ______________
Class ______________

OBJECTIVES:

PROCEDURE:

DATA:

Table 24–1

Primary winding resistance, R_P	
Secondary winding resistance, R_S	
Turns ratio, n (computed)	
Turns ratio, n (measured)	
% difference	

Observations from Step 5:

____

____

Observations from Step 6:

____

____

Table 24–2

Step	Condition	$V_{speaker}$
7	Output directly from generator	
8	With impedance-matching transformer	

RESULTS AND CONCLUSION:

FURTHER INVESTIGATION RESULTS:

APPLICATION PROBLEM RESULTS:

EVALUATION AND REVIEW QUESTIONS:

1. You measured the primary winding resistance and the secondary winding resistance and probably found that the primary resistance was much higher than the secondary resistance.

   (a) What factors do you think account for this?

   (b) Do you think a step-up transformer would have higher primary or secondary resistance? Explain your answer.

2. What factors might cause a difference between the measured and computed turns ratio in step 2 and step 3?

3. Compare the voltage across the speaker as measured in step 7 and in step 8. Explain why there is a difference.

4. The power that is supplied to an ideal transformer should be zero if there is no load. Why?

5. (a) If an ideal transformer has 115 V across the primary and draws 200 mA of current, what power is dissipated in the load?

   (b) If the secondary voltage in the transformer of part (a) is 24 V, what is the secondary current?

   (c) What is the turns ratio?

6. What is the reason that utility companies transmit power using very high voltages?

TECH TIP REVIEW:

Each of the following questions relate to the TECH TIP circuit shown in Figure 15–39 of the Floyd text. For these questions, assume that the transformer is ideal—that is, no power is dissipated in the transformer and the power in is equal to the power out.

1. Since the transformer shown in Figure 15–39 steps the voltage down, the current is stepped up. Assume the maximum output current is 1 A. What size fuse should be specified on the primary side to just limit the output current to 1 A?

2. Assume it is desired to use the low-voltage power supply in a location where the input ac voltage is 220 V ac instead of 110 V ac. The transformer is to be replaced with a different transformer to accommodate this change. The higher input voltage should have no effect on the output voltage or current.
   (a) What turns ratio should the new transformer have?

   (b) What fuse should be specified?

25 Series *RC* Circuits

PSpice

OBJECTIVES:

After performing this experiment, you will be able to:

1. Compute the capacitive reactance of a capacitor from voltage measurements in a series *RC* circuit.
2. Draw the impedance and voltage phasor diagrams for a series *RC* circuit.
3. Explain how frequency affects the impedance and voltage phasors in a series *RC* circuit.

READING:

Floyd, *Principles of Electric Circuits,* Sections 16–1 through 16–3

MATERIALS NEEDED:

One 6.8 kΩ resistor
One 0.01 μF capacitor
Application Problem: Capacitor, value to be determined by student

SUMMARY OF THEORY:

When a sine wave at some frequency drives a circuit that contains only linear elements (resistors, capacitors, and inductors), the waveforms throughout the circuit are also sine waves at that same frequency. To show the relationship between the sinusoidal voltages and currents, we can represent ac waveforms as phasor quantities. A *phasor* is a complex number that is used to represent a sine wave's amplitude and phase. A graphical representation of the phasors in a circuit is a useful tool for visualizing the amplitude and phase relationship of the various waveforms. The algebra of complex numbers can then be used to perform arithmetic operations on sine waves.

Figure 25–1(a) shows an *RC* circuit with its impedance phasor diagram plotted in Figure 25–1(b). The total impedance is 5 kΩ, producing a current of 1.0 mA. In any series circuit, the same current flows throughout the circuit. By multiplying each of the phasors in the impedance diagram by the current in the circuit, we arrive at the voltage phasor diagram illustrated in Figure 25–1(c). It is convenient to use current as the reference for comparing voltage phasors because it is the same throughout. Notice the direction of the current phasor. The voltage and the current are in the same direction across the resistor because they are in phase, but the voltage across the capacitor lags the current by 90°. The generator voltage is the phasor sum of the voltage across the resistor and the voltage across the capacitor.

The phasor diagram illustrated by Figure 25–1 is correct at only one frequency. This is because the reactance of a capacitor is frequency dependent as given by the equation

$$X_C = \frac{1}{2\pi fC}$$

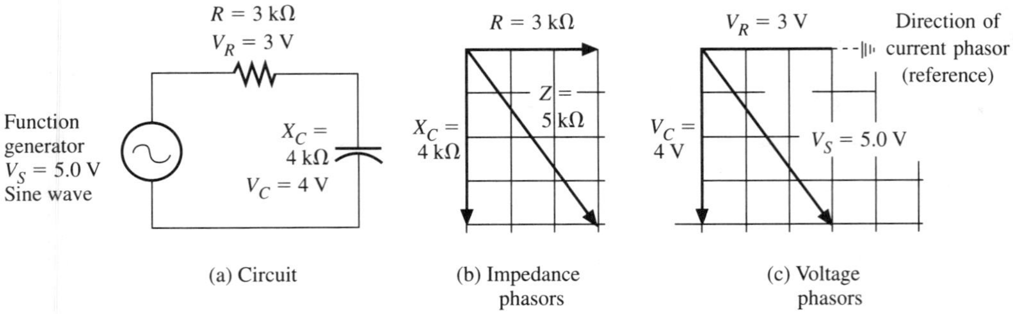

Figure 25–1

As the frequency is raised, the reactance (X_C) of the capacitor decreases. This changes the phase angle and voltages across the components. These changes will be investigated in this experiment.

PROCEDURE:

1. Measure the actual capacitance of a 0.01 μF capacitor and a 6.8 kΩ resistor. Enter the measured values in Table 25–1. If you cannot measure the capacitor, use the listed value.

2. Connect the series *RC* circuit shown in Figure 25–2. Set the function generator for a 500 Hz sine wave at 3.0 V_{pp}. The voltage should be measured with the circuit connected. Set the voltage with the oscilloscope and check both the voltage and the frequency with the scope. Record all voltages and currents throughout this experiment as peak-to-peak values.

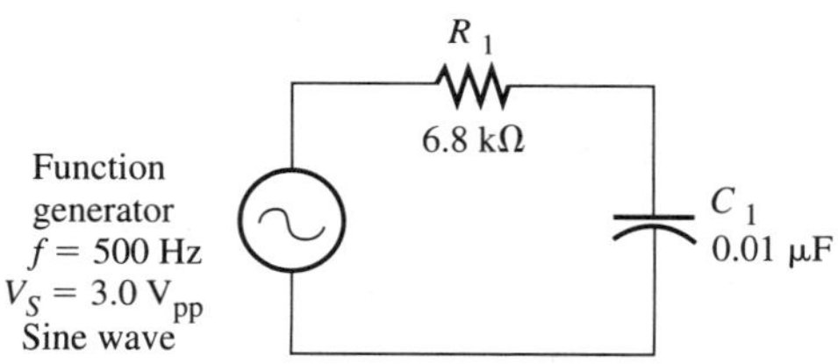

Figure 25–2

3. Using the two-channel-difference technique described in Experiment 18, measure the peak-to-peak voltage across the resistor (V_R). Then measure the peak-to-peak voltage across the capacitor (V_C). Record the voltage readings on the first line of Table 25–2.

4. Compute the peak-to-peak current in the circuit by applying Ohm's law to the measured value of the resistor:

$$I = \frac{V_R}{R}$$

Since the current is the same throughout a series circuit, this is a simple method for finding the current in both the resistor and the capacitor. Enter the computed current in Table 25–2.

5. Compute the capacitive reactance, X_C, by applying Ohm's law to the capacitor. The reactance is found by dividing the voltage across the capacitor (step 3) by the current in the circuit (step 4). Enter the capacitive reactance in Table 25–2.

6. Compute the total impedance of the circuit by applying Ohm's law to the entire circuit. Use the generator voltage set in step 2 and the current determined in step 4. Enter the computed impedance in Table 25–2.

7. Change the frequency of the generator to 1000 Hz. Check the generator voltage and reset it to 3.0 V_{pp} if necessary. Repeat steps 3 through 6, entering the data in Table 25–2. Continue in this manner for each frequency listed in Table 25–2.

8. The data in Table 25–2 indicate how the voltage across the resistor and the voltage across the capacitor vary with frequency. Plot both the voltage across the capacitor and the voltage across the resistor as a function of frequency on Plot 25–1 of your report.

9. From the data in Table 25–2 and the measured value of R_1, draw the impedance phasors for the circuit at a frequency of 2000 Hz on Plot 25–2(a) and the voltage phasors at the same frequency on Plot 25–2(b).

FOR FURTHER INVESTIGATION:

This experiment showed that the voltage phasor diagram can be obtained by multiplying each quantity on the impedance phasor diagram by the current in the circuit. In turn, if each of the voltage phasors is multiplied by the current, the resulting diagram is the power phasor diagram. Usually, power is reported with rms values. Using the data from Table 25–2, convert the current and the source voltage to an rms value. Then determine the true power, the reactive power, and the apparent power in the *RC* circuit at a frequency of 1000 Hz and at a frequency of 4000 Hz. On Plot 25–3, draw the power phasor diagrams. (See Section 16–7 of Floyd's text for further discussion of power phasors.)

APPLICATION PROBLEM:

If an amplifier is used only for ac signals, it is often useful to roll off the gain at dc. The circuit in Figure 25–3(a) shows how this can be done in an operational amplifier circuit using a *RC* series circuit. In this problem, R_B affects the gain and is required to be 6.8 kΩ. The desired roll-off frequency *($X_C = R$)* is 25 Hz. Compute the value of the capacitor that meets this specification. Construct the circuit shown in Figure 25–3(b) with your computed capacitor. Measure the frequency response and prove that the roll-off frequency meets the desired response.

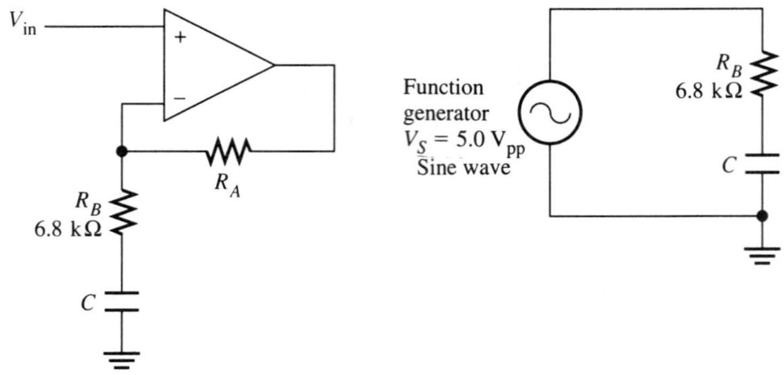

(a) Amplifier with low-frequency roll-off (b) Test circuit

Figure 25–3

PSpice EXAMPLE:

The following PSpice example allows you to readily plot the phasor diagram at a selected frequency for the circuit in Figure 25–2. The printout will show the voltage, current, and phase angle for the components as a function of frequency.

```
LAB 25 FIG 25–2
VS 1 0 AC 3V
R1 1 2 6.8E3
C1 2 0 .01E–6
.AC LIN 25 500Hz 8000Hz
.PRINT AC V(R1) VP(R1) I(R1) IP(R1) V(C1) VP(C1) I(C1) IP(C1)
.PROBE
.OPTIONS NOPAGE
.END
```

Report for Experiment 25

Name ____________________
Date ____________________
Class ____________________

OBJECTIVES:

PROCEDURE:

DATA:

Table 25–1

Component	Listed Value	Measured Value
C_1	0.01 μF	
R_1	6.8 kΩ	

Table 25–2

Frequency	V_R	V_C	I	X_C	Z
500 Hz					
1000 Hz					
1500 Hz					
2000 Hz					
4000 Hz					
8000 Hz					

Step 8 (Graph of Voltage Versus Frequency):

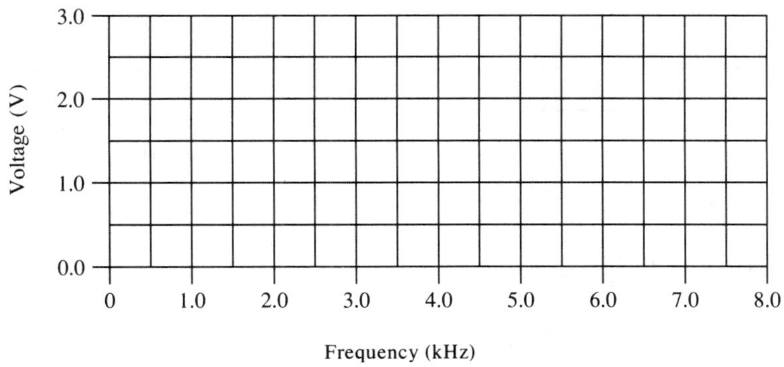

Plot 25–1

Step 9 (Impedance and Voltage Phasors for 2000 Hz):

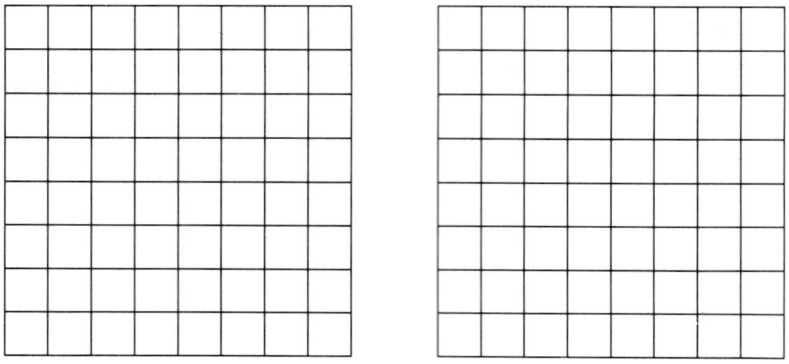

Plot 25–2

RESULTS AND CONCLUSION:

FURTHER INVESTIGATION RESULTS:

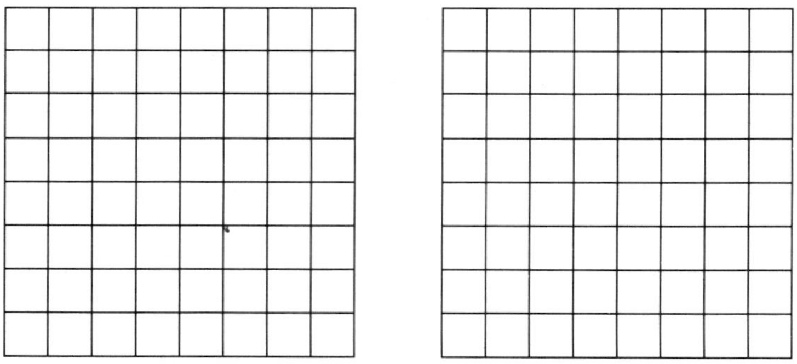

f = 1000 Hz　　　　*f* = 4000 Hz

Plot 25–3

APPLICATION PROBLEM RESULTS:

EVALUATION AND REVIEW QUESTIONS:

1. The Pythagorean theorem can be applied to the phasors drawn in Plots 25–2. Show that the data in both plots satisfy the equations

$$Z = \sqrt{R^2 + X_C^2}$$

$$V_S = \sqrt{V_R^2 + V_C^2}$$

2. Assume you needed to pass high frequencies through an *RC* filter but block low frequencies. From the data in Plot 25–1, should you connect the output across the capacitor or across the resistor? Explain your answer.

3. (a) What happens to the total impedance of a series *RC* circuit as the frequency is increased?

   (b) Explain why the phase angle between the generator voltage and the resistor voltage decreases as the frequency is increased:

4. A student accidentally used a capacitor that was ten times larger than required in the experiment. Predict what happens to the frequency response shown in Plot 25–1 with the larger capacitor.

5. Assume no current flowed in the series *RC* circuit because of an open circuit. How could you quickly determine if the resistor or the capacitor were open?

6. From Plot 25–1, predict the frequency at which the phase shift is 45° between the source voltage and current.

26 Parallel *RC* Circuits

PSpice

OBJECTIVES:

After performing this experiment, you will be able to:

1. Measure the current phasors for a parallel *RC* circuit.
2. Explain how the current phasors and phase angle are affected by a change in frequency for parallel *RC* circuits.

READING:

Floyd, *Principles of Electric Circuits,* Sections 16–4 through 16–5 and 16–10

MATERIALS NEEDED:

One 100 kΩ resistor
Two 1.0 kΩ resistors
One 1000 pF capacitor
Application Problem: One 10 kΩ potentiometer, one 100 kΩ resistor, one 0.1 μF capacitor

SUMMARY OF THEORY:

In a series circuit, the same *current* is in all components. For this reason, current is used as a reference. In addition, Kirchhoff's voltage law applies to reactive circuits provided the voltages are added as phasors. By contrast, in parallel, the same *voltage* is across all components. The voltage is, therefore, the reference. Current in each branch is compared to the circuit voltage. In parallel circuits, Kirchhoff's current law applies to any junction. Current entering a junction must be equal to current leaving the junction. Again, care must be taken to add the currents as phasors.

Figure 26–1 illustrates a parallel *RC* circuit. If the impedance of each branch is known, the current in that branch can be determined directly from Ohm's law. The current phasor diagram can then be constructed directly. The total current can be found as the phasor sum of the currents in each branch. The current in the capacitor is shown at +90° from the voltage reference because the current leads the voltage in a capacitor. The current in the resistor is along the *x*-axis because current and voltage are in phase in a resistor. The Pythagorean theorem can be applied to the current phasors, resulting in the equation

$$I_T = \sqrt{I_R^2 + I_C^2}$$

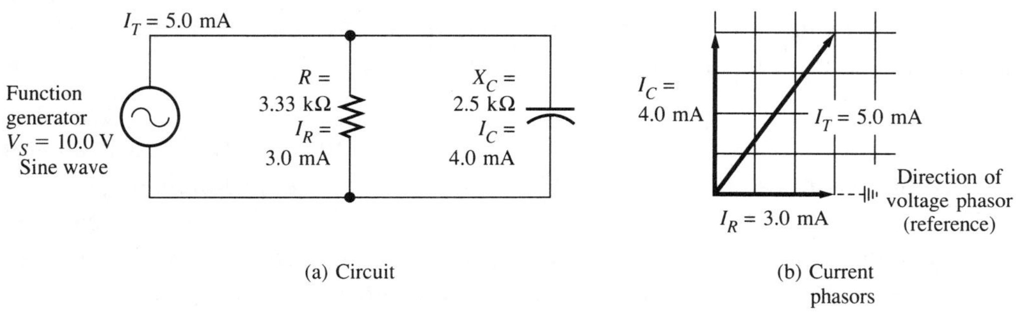

Figure 26–1

In this experiment, two extra 1.0 kΩ resistors are added to "sense" current and provide a small voltage drop that can be measured. These resistors are much smaller than the parallel branch impedance, so their resistance can be ignored in the computation of circuit impedance.

PROCEDURE:

1. Measure a resistor with a color-coded value of 100 kΩ and each of two current sense resistors (R_{S1} and R_{S2}) with color-coded values of 1.0 kΩ. Measure the capacitance of a 1000 pF capacitor. Use the listed value if a measurement cannot be made. Record the measured values in Table 26–1.

2. Construct the circuit shown in Figure 26–2. Because you will be measuring some very small voltages in this experiment, use a voltmeter and record all voltages and currents as rms values. Set the generator to a voltage of 3.0 V_{rms} at 1.0 kHz. Check both the voltage and frequency with your oscilloscope. (Note that 3.0 V_{rms} is approximately 8.5 V_{pp}.)

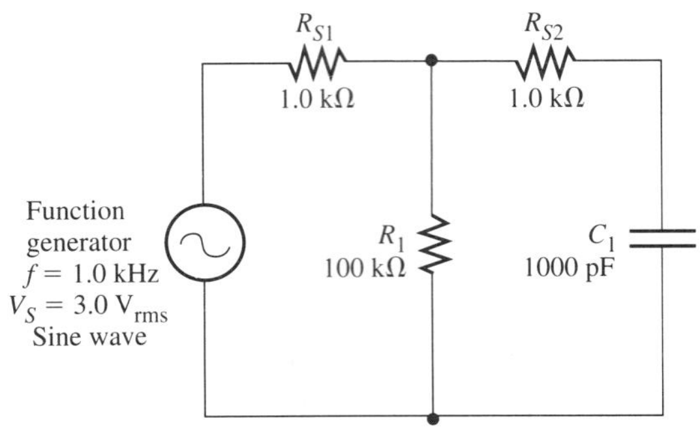

Figure 26–2

3. Using the voltmeter, measure the rms voltage drop across each resistor. The voltage drops across the sense resistors are particularly small, so measure them as accurately as possible, keeping three significant figures in your answer. Record the voltage drops in Table 26–1.

4. Compute the rms current in each resistor using Ohm's law. Record the computed current in Table 26–1.

5. Draw the current phasors I_{R1} and I_{C1} and the total current I_T on Plot 26–1. The total current flows through sense resistor R_{S1}. The current I_{C1} flows through sense resistor R_{S2}. Ignore the small effect of the sense resistors on the phasor diagram. Note carefully the direction of the phasors. Label each of the current phasors.

6. Compute X_{C1} for the 1.0 kHz frequency. Using this value and the measured resistance of R_1, compute the total impedance of the circuit using the product-over-sum rule for parallel resistance and reactance (see Floyd's text). The sense resistors can be ignored for this calculation, but you will need to use phasor math to obtain the correct result. Record the values of X_{C1} and Z_T in the space provided in the report.

7. Using the Z_T found in step 6 and the applied voltage, V_S, compute the total current, I_T. Check that the total current agrees within experimental error with the value determined in step 4. Record the computed current in the space provided in the report.

8. Change the frequency of the generator to 2.0 kHz. Check that generator voltage is still 3.0 V_{rms}. Repeat steps 1–5 for the 2.0 kHz frequency.* Verify all measurements with the oscilloscope. Enter the data in Table 26–2 and draw the current phasors on Plot 26–2.

FOR FURTHER INVESTIGATION:

In a series *RC* circuit, the impedance phasor is the sum of the resistance and reactance phasors, as shown in Experiment 22. In a parallel circuit, the admittance phasor is the sum of the conductance and the susceptance phasors. On Plot 26–3, draw the admittance, conductance, and susceptance phasors for the experiment at a frequency of 1.0 kHz. (*Hint:* The admittance phasor diagram can be obtained directly from the current phasor diagram by dividing the current phasors by the applied voltage.)

APPLICATION PROBLEM:

Tone-control circuits can be used to limit the band of frequencies in an audio system to limit high- or low-frequency noise. The tone-control circuit shown in Figure 26–3 uses both series and parallel *RC* circuits to provide high- and low-frequency attenuation. Construct the circuit. Connect one channel of your oscilloscope across the input signal generator and the other channel across the load resistor and view the signals together. Start by setting the signal generator to 1 V_{pp} at 100 Hz. Which control affects the low-frequency response? Try setting the generator to 1 kHz and then to 10 kHz. Test the effect of each control on the response. Explain the circuit and your measurements in your report.

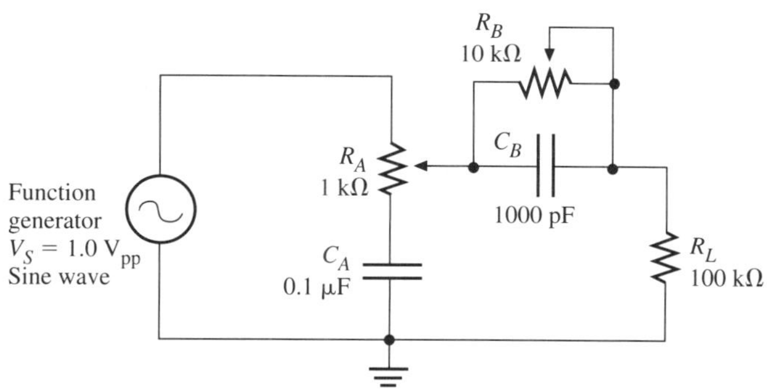

Figure 26–3

*Most DMMs can measure the relative voltages at this frequency; however, some DMMs will not give accurate absolute voltages. Relative voltages will allow impedances to be correctly determined.

PSpice EXAMPLE:

The following example will compute the voltages and currents for Figure 26–2 at 1 kHz.

```
LAB26 FIG 26-2
VS 1 0 AC 3
RS1 1 2 1E3
R1 2 0 10E3
RS2 2 3 1E3
C1 3 0 1000E-12
.AC LIN 1 1E3Hz 1E3Hz
.PRINT AC V(R1) VP(R1) V(C1) VP(C1) I(R1) IP(R1) I(C1) IP(C1)
.OPTIONS NOPAGE
.END
```

Report for Experiment 26

Name ________________
Date ________________
Class ________________

OBJECTIVES:

PROCEDURE:

DATA:

Table 26–1 (f = 1.0 kHz)

	Listed Value	Measured Value	Voltage Drop	Computed Current
R_1	100 kΩ			
R_{S1}	1.0 kΩ			
R_{S2}	1.0 kΩ			
C_1	1000 pF			

X_{C1} = ____ Z_T = ____

I_T = ____

Plot 26–1

Table 26–2 (f = 2.0 kHz)

	Listed Value	Measured Value	Voltage Drop	Computed Current
R_1	100 kΩ			
R_{S1}	1.0 kΩ			
R_{S2}	1.0 kΩ			
C_1	1000 pF			

X_{C1} = ____ Z_T = ____

I_T = ____

Plot 26–2

RESULTS AND CONCLUSION:

FURTHER INVESTIGATION RESULTS:

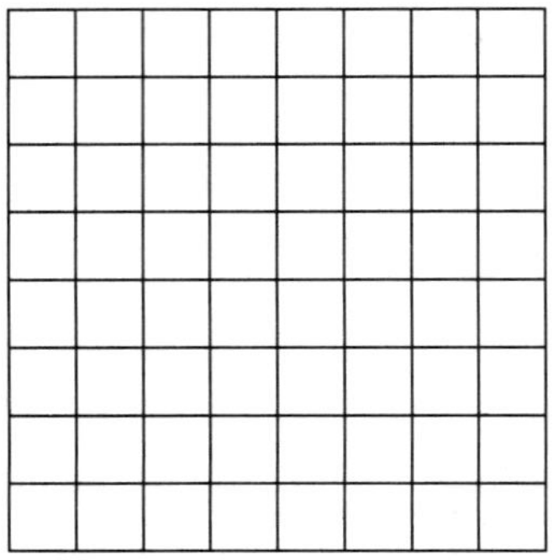

Plot 26–3

APPLICATION PROBLEM RESULTS:

EVALUATION AND REVIEW QUESTIONS:

1. Explain how increasing the frequency affects each of the following.
   (a) the total impedance of the circuit

   (b) the phase angle between the generator voltage and the generator current

2. Assume the frequency had been set to 5.0 kHz in this experiment. Compute:
   (a) the current in the resistor ____________
   (b) the current in the capacitor ____________
   (c) the total current ____________

3. If a smaller capacitor had been substituted in the experiment, what would happen to the current phasor diagrams?

4. (a) A common circuit using a parallel *RC* circuit is the bypass capacitor across the emitter resistor in a transistor amplifier, as illustrated in Figure 26-4. The frequency at which the resistance R_E is equal to the capacitive reactance X_C is called the *cutoff* frequency. Compute the cutoff frequency for the circuit by setting $R_E = X_C$ and solving for *f*.

   $f =$ ______

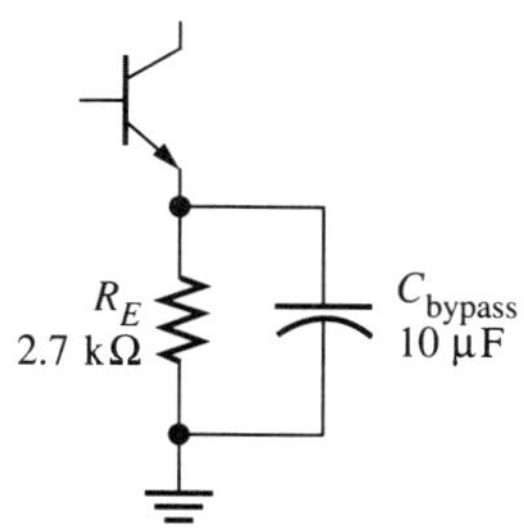

Figure 26–4

   (b) How do the branch currents compare at the cutoff frequency?

   (c) Explain what happens above this frequency to the current in the parallel *RC* circuit.

5. If the capacitance in Figure 26–4 is increased, what happens to the cutoff frequency?

6. What is the phase angle between the source voltage and current at the cutoff frequency?

TECH TIP REVIEW:

Each of the following questions relate to the TECH TIP circuit shown in Figure 16–64 of the Floyd text. Notice that the circuit drawn in Figure 16–64 has been redrawn as an equivalent series *RC* circuit in Figure 16–65(c) by combining the 47 kΩ and 10 kΩ resistors.

1. In Experiment 25, you observed the response of a series *RC* circuit. From your experimental measurements, what do you expect to see at point B in the circuit as the frequency of the ac source is reduced?

2. At a certain frequency, called the cutoff frequency, the reactance of the capacitor is equal to the resistance of the equivalent resistor in Figure 16–65(c). At the cutoff frequency, 70.7% of the input voltage is across the capacitor and 70.7% of the input voltage is across the equivalent resistor.

   (a) How do you reconcile these facts against Kirchhoff's voltage law?

   (b) How would you change the capacitor's value if you wanted to have a lower cutoff frequency?

27 Series *RL* Circuits

OBJECTIVES:

After performing this experiment, you will be able to:

1. Compute the inductive reactance of an inductor from voltage measurements in a series *RL* circuit.
2. Draw the impedance and voltage phasor diagrams for the series *RL* circuit.
3. Measure the phase angle in a series circuit using either of two methods.

READING:

Floyd, *Principles of Electric Circuits,* Sections 17–1 through 17–3

MATERIALS NEEDED:

One 10 kΩ resistor
One 100 mH inductor
Application Problem: Compass, protractor, ruler

SUMMARY OF THEORY:

When a sine wave drives a linear series circuit, the phase relationships between the current and the voltage are determined by the components in the circuit. The current and voltage are always in phase across resistors. In capacitors, the current is always leading the voltage by 90°, but for inductors, the voltage always leads the current by 90°. (A simple memory aid for this is *ELI the ICE man,* where *E* stands for voltage, *I* for current, and *L* and *C* for inductance and capacitance.)

Figure 27–1(a) illustrates a series *RL* circuit. The graphical representation of the phasors for this circuit is shown in Figure 27–1(b) and (c). As in the series *RC* circuit, the total impedance is obtained by adding the resistance and inductive reactance using the algebra for complex numbers. In this example, the current is 1.0 mA and the total impedance is 5.0 kΩ. The current is the same in all components of a series circuit, so the current is drawn as a reference in the direction of the *x*-axis. If the current is multiplied by the impedance phasors, the voltage phasors are obtained as shown in Figure 27–1(c).

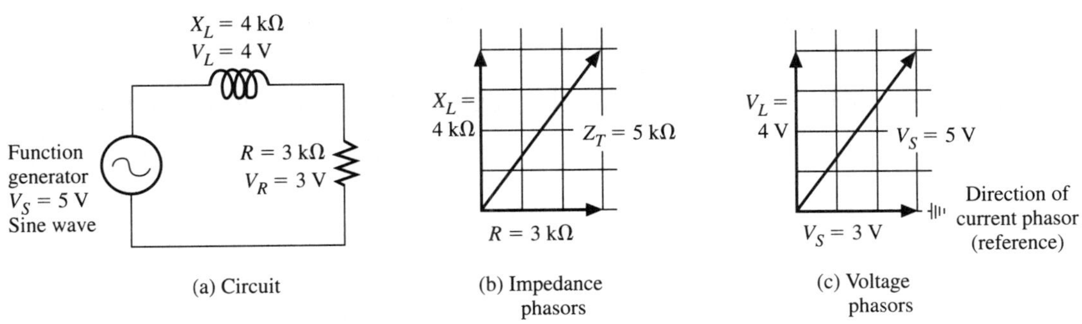

Figure 27–1

In this experiment, you will learn how to make measurements of the phase angle. Actual inductors may have enough resistance to affect the phase angle in the circuit. To avoid this error, you will use a series resistor that is large compared to the inductor's resistance.

PROCEDURE:

1. Measure the actual resistance of a 10 kΩ resistor and the inductance of a 100 mH inductor. If the inductor cannot be measured, record the listed value. Record the measured values in Table 27–1.

2. Connect the circuit shown in Figure 27–2. Set the function generator voltage with the circuit connected to 3.0 V_{pp} at a frequency of 25 kHz. The generator should have no dc offset. Measure the generator voltage and frequency with the oscilloscope, since most meters cannot respond to the 25 kHz frequency. Use peak-to-peak readings for all voltage and current measurements in this experiment.

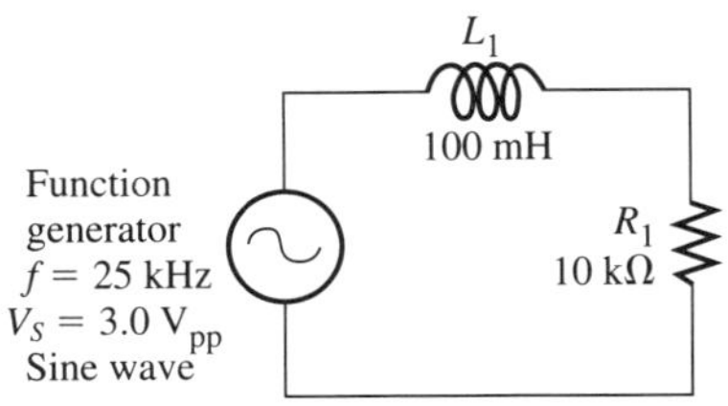

Figure 27–2

3. Using a two-channel oscilloscope, measure the peak-to-peak voltage across the resistor (V_R) and the peak-to-peak voltage across the inductor (V_L). (See Figure 27–3 for setup.) Measure the voltage across the inductor using the difference technique described in Experiment 18. Record the voltage readings in Table 27–2.

4. Compute the peak-to-peak current in the circuit by applying Ohm's law to the resistor; that is,

$$I = \frac{V_R}{R}$$

Enter the computed current in Table 27–2.

5. Compute the inductive reactance, X_L, by applying Ohm's law to the inductor. The reactance is

$$X_L = \frac{V_L}{I}$$

Enter the computed reactance in Table 27–2.

6. Calculate the total impedance (Z_T) by applying Ohm's law to the entire circuit. Use the generator voltage set in step 2 (V_S) and the current determined in step 4. Enter the computed impedance in Table 27–2.

7. Using the values listed in Table 27–1 and Table 27–2, draw the impedance phasors in your report on Plot 27–1(a) and the voltage phasors on Plot 27–1(b) at a frequency of 25 kHz.

8. Compute the phase angle between V_R and V_S using the trigonometric relation

$$\theta = \tan^{-1}\left(\frac{V_L}{V_R}\right)$$

Enter the computed phase angle in Table 27–3.

9. Two methods for measuring phase angle will be explained. Measure the phase angle between V_R and V_S using each method. The measured phase angle will be recorded in Table 27–3.

Phase Angle Measurement—Method 1:

(a) Connect the oscilloscope so that channel 1 is across the generator and channel 2 is across the resistor. (See Figure 27–3.) Obtain a stable display showing a little more than one cycle while viewing channel 1 (V_S). The scope should *not* be triggered using composite triggering.

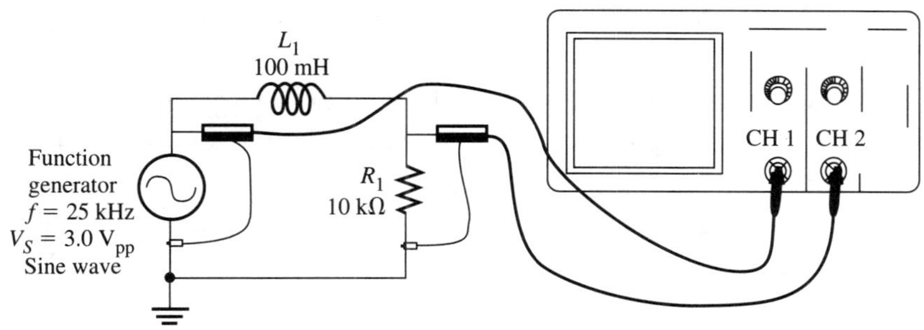

Figure 27–3

(b) Measure the period, *T*, of the generator. Record the result in Table 27–3. You will use this in step (e).

(c) Set the oscilloscope to view both channels. Adjust the amplitudes of the signals using the VOLTS/DIV, VERT POSITION, and red vernier controls until both channels appear to have the same amplitude, as seen on the scope face.

(d) Spread the signal horizontally using the SEC/DIV control until both signals are just visible across the screen. The SEC/DIV control must remain calibrated. Measure the time between the two signals, Δt, by counting the number of divisions along a horizontal graticule of the oscilloscope and multiplying by the SEC/DIV setting. (See Figure 27–4.) Record the measured Δ*t* in Table 27–2.

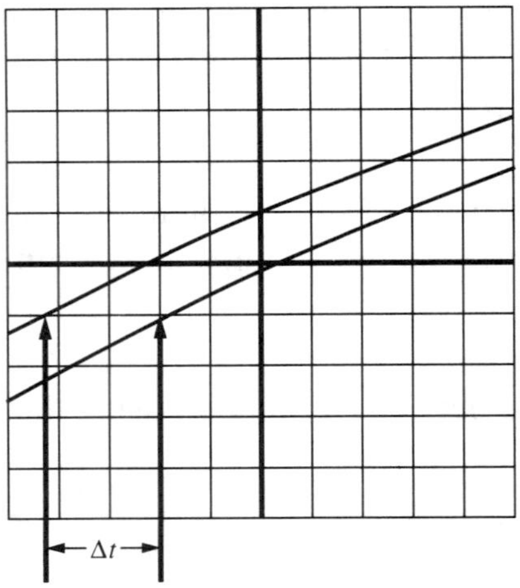

Figure 27–4

(e) The phase angle may now be computed from the equation

$$\Theta = \left(\frac{\Delta t}{T}\right) \times 360^\circ$$

Enter the measured phase angle in Table 27–3 as Phase Angle: Method 1.

Phase Angle Measurement—Method 2:

(a) In this method the oscilloscope face represents degrees and the phase angle can be measured directly. The probes are connected as before. View channel 1 and obtain a stable display. Then adjust the SEC/DIV control and its vernier until you have exactly one cycle across the scope face. This is equivalent to 360° in 10 divisions, so each division is worth 36°.*

(b) Now switch the scope to view both channels. As before, adjust the amplitudes of the signals using the VOLTS/DIV, VERT POSITION, and red vernier controls until both channels appear to have the same amplitude.

(c) Measure the number of divisions between the signals and multiply by 36° per division. Record the measured phase angle in Table 27–3 as Phase Angle: Method 2.

FOR FURTHER INVESTIGATION:

An older method for measuring phase angles involved interpreting Lissajous figures. A Lissajous figure is the pattern formed by the application of a sinusoidal waveform to both the *x*- and *y*-axes of an oscilloscope. Two signals of equal amplitude and exactly in phase will produce a 45° line on the scope face. If the signals are the same amplitude and exactly 90° apart, the waveform will appear as a circle. Other phase angles can be determined by applying the formula

$$\Theta = \sin^{-1}\frac{\mathbf{OA}}{\mathbf{OB}}$$

*Alternatively, you can set one-half cycle across the screen, making each division worth 18°. (See Floyd's TECH TIP #12.)

Figure 27–5 illustrates a Lissajous figure phase measurement. The measurement of **OA** and **OB** is along the *y*-axis. Try measuring the phase angle in this experiment using a Lissajous figure. The signals will have to have the same amplitude and be centered on the oscilloscope face. Then switch the time base of the oscilloscope to the X-Y mode.

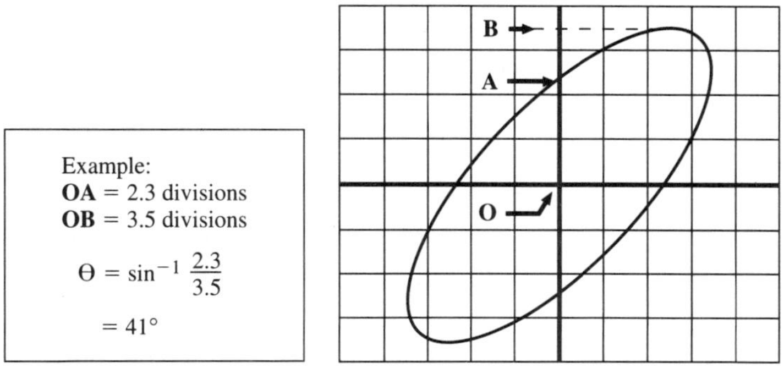

Figure 27–5

APPLICATION PROBLEM:

The phasor drawings illustrated so far can be used as a graphical solution for problems with reactive circuits operating at some specified frequency. Another graphical solution to a problem is illustrated using a *circle diagram*. This method plots the current in the circuit as one variable (in this case the inductive reactance) changes.

Consider the *RL* circuit shown in Figure 27–1(a). Applying Ohm's law, the current in the circuit is 1.0 mA and lags the voltage by 53.1°. This phasor is plotted in Figure 27–6(a). If the frequency is varied, the phasor drawing changes due to the change in the inductive reactance, X_L. At dc, the inductive reactance is zero (assuming an ideal inductor) and the current in the circuit is determined only by the resistance. In this case, the current is 1.67 mA at an angle of zero (5 V across 3 kΩ). This phasor is also plotted in Figure 27–6(a).

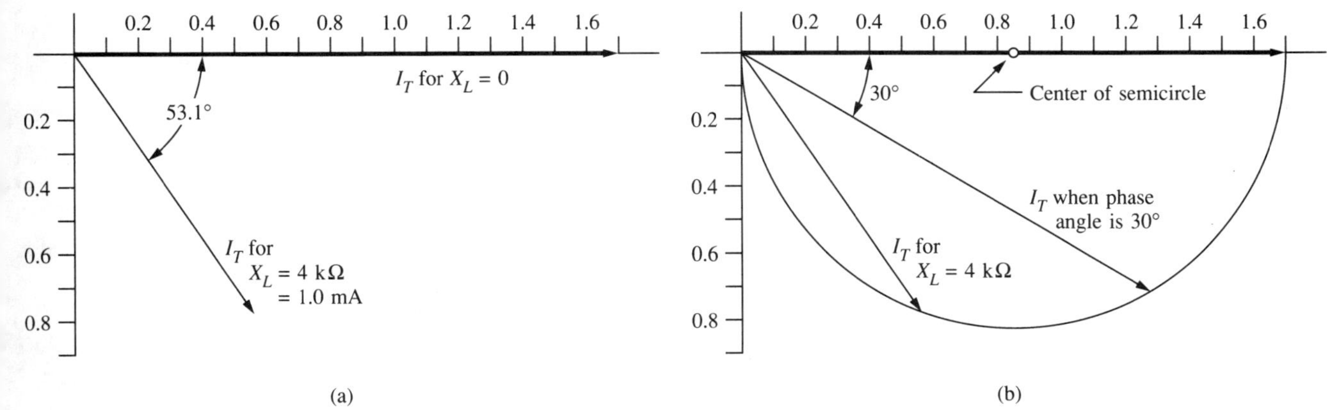

Figure 27–6

If you compute the current in the circuit as you vary the frequency, the tips of the current phasors would plot a semicircle, as shown in Figure 27–6(b). The center of the semicircle is in the middle of the resistance current phasor. This graph makes it easy to answer questions such as, What is the current in the circuit when the current lags the voltage by 30°? The answer to this question is plotted in Figure 27–6(b).

For this application problem, plot the current phasor for the experiment on a circle diagram. Then, from your graph, find the current in the circuit when the phase angle is 60°. Show the current phasor on your solution.

Report for Experiment 27

Name ____________________
Date ____________________
Class ____________________

OBJECTIVES:

PROCEDURE:

DATA:

Table 27–1

Component	Listed Value	Measured Value
L_1	100 mH	
R_1	10 kΩ	

Table 27–2
(f = 25 kHz)

V_R	V_L	I	X_L	Z_T

Step 7 (Phasor Drawings):

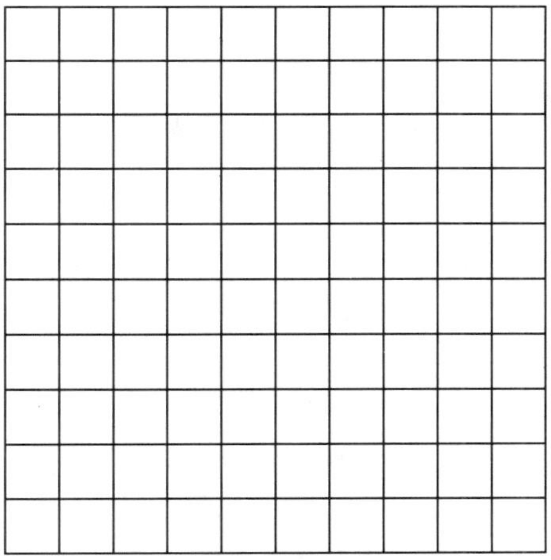

(a)

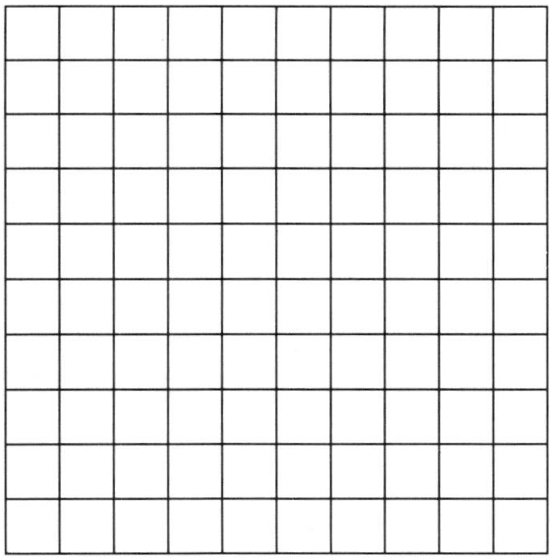

(b)

Plot 27–1

Table 27–3

Computed Phase Angle Θ	Measured Period T	Time Difference Δt	Phase Angle	
			Method 1 Θ	Method 2 Θ

RESULTS AND CONCLUSIONS:

FURTHER INVESTIGATION RESULTS:

APPLICATION PROBLEM RESULTS:

EVALUATION AND REVIEW QUESTIONS:

1. (a) What will happen to the impedance in this experiment if the frequency increases?

   (b) What will happen to the impedance if the inductance is larger?

2. (a) What will happen to the phase angle in this experiment if the frequency increases?

   (b) What will happen to the phase angle if the inductance is larger?

3. (a) Compute the percent difference between the computed phase angle and the method 1 phase angle measurement.

   (b) Compute the percent difference between the computed phase angle and the method 2 measurement.

   (c) Which method was most accurate?

4. The method 2 phase measurement directed you to set up the waveform so that each division across the oscilloscope was equal to 36°. Explain how you could use this idea to set up each division for 18°. What precaution must be observed for this setup?

5. (a) The critical frequency for an *RL* circuit occurs at the frequency at which the resistance is equal to the inductive reactance. That is, $R = X_L$. Since $X_L = 2\pi fL$ for an inductor, it can be easily shown that the critical frequency for an *RL* circuit is

$$f_{\text{CRIT}} = \frac{R}{2\pi L}$$

Compute the critical frequency for this experiment. What is the phase between V_R and V_S at the critical frequency?

f_{CRIT} = ____________ θ = ____________

6. A series *RL* circuit contains a 100 Ω resistor and a 1.0 H inductor and is operating at a frequency of 60 Hz. If 3.0 V are across the resistor, compute:

(a) the current in the inductor ____________

(b) the inductive reactance X_L ____________

(c) the voltage across the inductor, V_L ____________

(d) the source voltage, V_S ____________

(e) the phase angle between V_R and V_S ____________

28 Parallel *RL* Circuits

OBJECTIVES:

After performing this experiment, you will be able to:

1. Determine the current phasor diagram for a parallel *RL* circuit.
2. Measure the phase angle between the current and voltage for a parallel *RL* circuit.
3. Explain why an inductive circuit differs from the ideal model.

READING:

Floyd, *Principles of Electric Circuits,* Sections 17–4 through 17–10

MATERIALS NEEDED:

One 3.3 kΩ resistor
Two 47 Ω resistors
One 100 mH inductor

SUMMARY OF THEORY:

Parallel *RC* circuits were investigated in Experiment 26. Recall that the circuit phasor diagram was drawn with current phasors and the voltage phasor was used as a reference since voltage is the same across parallel components. In a parallel *RL* circuit, the current phasors will again be drawn with reference to the voltage phasor. The direction of the current phasor in a resistor is always in the direction of the voltage. Since current lags the voltage in an inductor, the current phasor is drawn at an angle of $-90°$ from the voltage reference. An example of a parallel *RL* circuit and the associated phasors are shown in Figure 28–1.

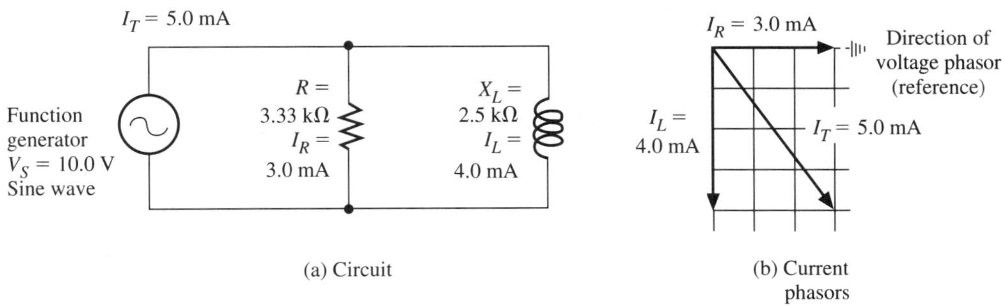

Figure 28–1

Practical inductors contain resistance that is large enough to affect the purely reactive inductive phasor drawn in Figure 28–1. The resistance of an inductor can be thought of as a resistor in series with a pure inductor. The effect on the phasor diagram is to reduce the angle between I_L and I_R. In a practical circuit this angle will be less than 90° indicated in Figure 28–1. This experiment will illustrate the difference between the approximations of circuit performance based on ideal components and the actual measured values.

Recall that in Experiment 27, the phase angle between the source voltage, V_S, and the resistor voltage, V_R, in a series circuit was measured. The oscilloscope is a voltage-sensitive device, so comparing these voltages is straightforward. In parallel circuits, the phase angle of interest is usually between the total current, I_T, and one of the branch currents. To use the oscilloscope to measure the phase angle in a parallel circuit, we must convert each branch current to a voltage. This is done by inserting a small resistor in the branch where the current is to be measured. The resistor must be small enough to not have a major effect on the circuit. In Experiment 26, you used sense resistors to measure the magnitude of the currents, but you computed rather than measured the phase angle. In this experiment, you will use the oscilloscope to measure both the magnitude and the phase angle between the source voltage and current.

PROCEDURE:

1. Measure the actual resistance of a resistor with a color-coded value of 3.3 kΩ and the resistance of two current-sensing resistors of 47 Ω each. Measure the inductance of 100 mH inductor. Use the listed value if you cannot measure the inductor. Record the measured values in Table 28–1.

2. Measure the coil resistance of the inductor with an ohmmeter. Record the resistance in Table 28–1.

3. Construct the circuit shown in Figure 28–2. Notice that the reference ground connection is at the low side of the generator. This connection will enable you to use a generator that does not have a "floating" common connection. Using your oscilloscope, set the function generator to a voltage of 6.0 V_{pp} sine wave at 5.0 kHz. Check both the voltage and frequency with your oscilloscope. Record all voltages and currents in this experiment as peak-to-peak values.

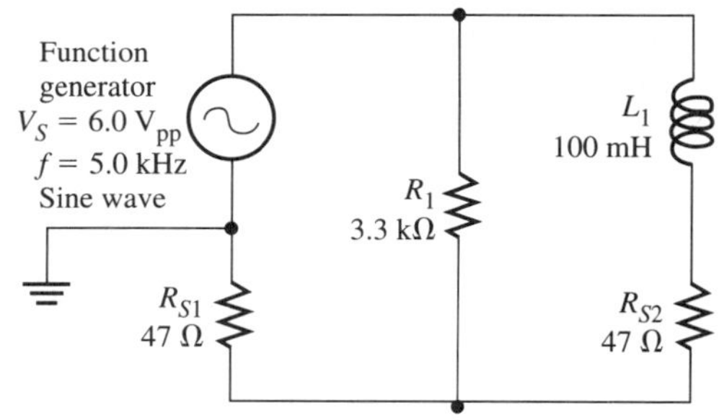

Figure 28–2

4. Using the oscilloscope, measure the peak-to-peak voltages across R_1, R_{S1}, and R_{S2}. Use the two-channel difference method (described in Experiment 18) to measure the voltage across the ungrounded components. Apply Ohm's law to the compute the current in each branch. Record the measured voltage drops and the computed currents in Table 28–1. Since L_1 is in series with R_{S2}, enter the same current for both.

5. The currents measured indirectly in step 4 are phasors because the current in the inductor is lagging the current in R_1 by 90°. The current in the inductor is the same as the current in R_{S2}, and the total current flows through R_{S1}. Using the computed peak-to-peak currents from Table 28–1, draw the current phasors for the circuit on Plot 28–1. (Ignore the effects of the sense resistors.)

6. The phasor diagram illustrates the relationship between the total current and the current in each branch. Using the measured currents, compute the phase angle between the total current (I_T) and the current in R_1 (I_R). Also compute the phase angle between the total current (I_T) and the current in the inductor (I_L). Enter the computed currents in Table 28–2.

7. In this step, you will measure the phase angle between the generator voltage and current. This angle is nearly identical to the angle between I_T and I_R, as shown in Figure 28–1. (Why?) Connect the oscilloscope probes as shown in Figure 28–3. Measure the phase angle using one of the methods in Experiment 27. The signal levels in each channel are quite different, so the vertical sensitivity controls should be adjusted to make each signal appear to have the same amplitude on the scope face. Record the measured result as the angle between I_T and I_R in Table 28–2.

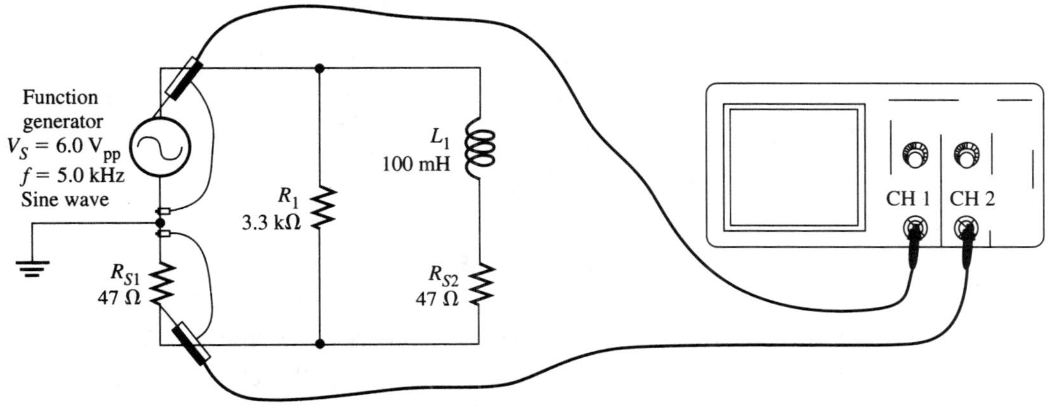

Figure 28–3

8. Replace R_{S1} with a jumper. This procedure enables you to reference the low side of R_1 and R_{S2}. Measure the angle between I_L and I_R by connecting the probes as shown in Figure 28–4. Ideally, this measurement should be 90°, but because of the coil resistance, you will likely find a smaller value. Adjust both channels for the same apparent amplitude on the scope face. Record your measured result in the second line in Table 28–2.

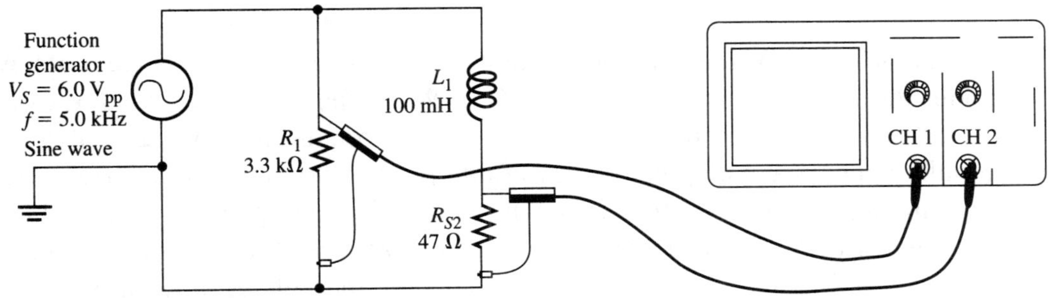

Figure 28–4

9. By subtracting the angle measured in step 7 from the angle measured in step 8, you can find the phase angle between the I_T and I_L. Record this as the measured value on the third line of Table 28–2.

FOR FURTHER INVESTIGATION:

We could find the *magnitude* of the total current by observing the loading effect of the circuit on the function generator. Consider the function generator as a Thevenin circuit consisting of a zero-impedance generator driving an internal series resistor consisting of the Thevenin source impedance. (See Figure 28–5.) When current flows to the external circuit, there is a voltage drop across the Thevenin resistance. The voltage drop across the Thevenin resistor, when divided by the Thevenin resistance, represents the total current in the circuit. To find the voltage drop across the Thevenin resistance, simply measure the difference in the generator voltage with the generator connected and disconnected from the circuit.

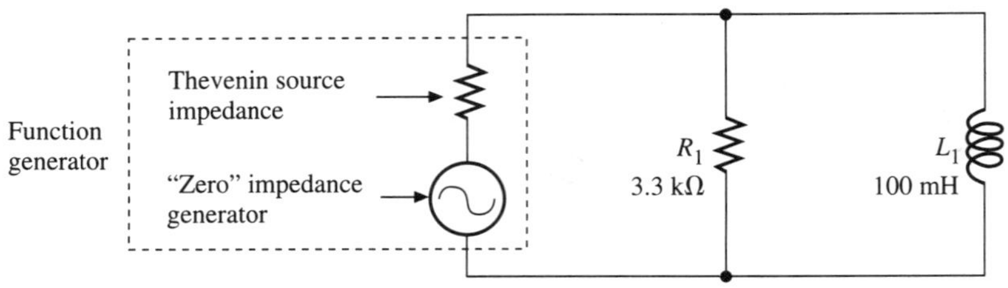

Figure 28–5

In addition to loading effects, the generator impedance also changes the phase angles in the circuit which is connected to it. If the impedance is smaller, the effect is greater. Investigate the loading effects for this experiment. Try finding the total current by the difference in loaded and unloaded voltage from the generator. What effect does the generator's impedance have on the phase angle?

APPLICATION PROBLEM:

The fundamental relationship for voltage versus current in an inductor is given by Lenz's law:

$$V_{\text{ind}} = L\left(\frac{di}{dt}\right)$$

This law is the basis for the statement that an inductor opposes a change in its own current. Consider the circuit shown in Figure 28–6(a). When the switch is closed, the current in the coil cannot change instantaneously. Since the current through the coil begins at zero, the inductance initially appears as an open circuit (infinite resistance). When the current reaches a steady state, the inductance appears to be a small resistor containing only its internal coil resistance in parallel with R. Construct the circuit using a square wave generator in place of the switch as shown in Figure 28–6(b). Observe the current and voltage waveforms for the inductor using a square wave from the signal generator. The current in the coil has the same shape as the voltage across the sense resistor, so you can monitor the current by connecting one channel of your oscilloscope across the sense resistor, R_S. Sketch the current and voltage waveforms for the coil. How do the observed waveforms show the application of Lenz's law?

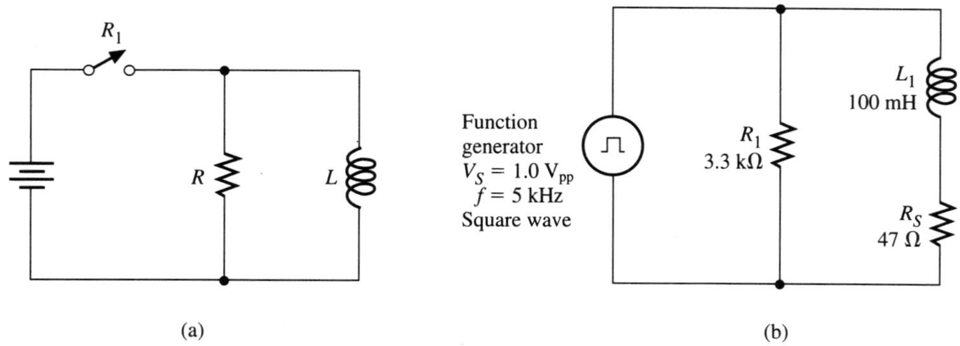

R1
R
L
(a)
Function
generator
VS = 1.0 Vpp
f = 5 kHz
Square wave
R1
3.3 kΩ
L1
100 mH
RS
47 Ω
(b)

Figure 28–6

Report for Experiment 28

Name ______________________
Date ______________________
Class ______________________

OBJECTIVES:

PROCEDURE:

DATA:

Table 28–1

	Listed Value	Measured Value	Voltage Drop	Computed Current
R_1	3.3 kΩ			
R_{S1}	47 Ω			
R_{S2}	47 Ω			
L_1	100 mH			
L_1 resistance				

Table 28–2

Phase Angle between:	Computed	Measured
I_T and I_R		
I_R and I_L	90°	
I_T and I_L		

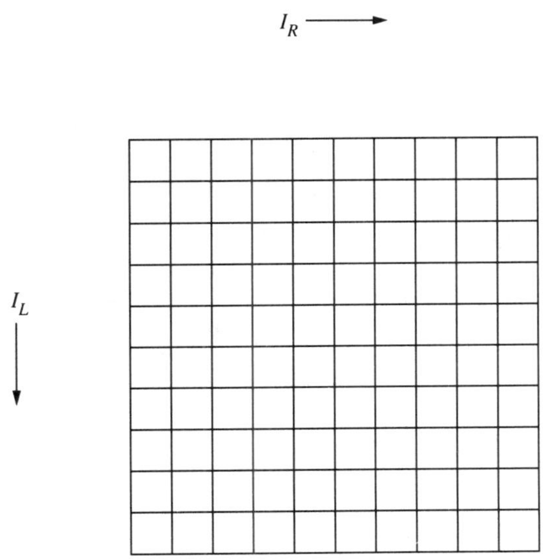

Plot 28–1

RESULTS AND CONCLUSION:

FURTHER INVESTIGATION RESULTS:

APPLICATION PROBLEM RESULTS:

EVALUATION AND REVIEW QUESTIONS:

1. The Pythagorean theorem is based on a right triangle—it assumes that the currents determined in step 4 were 90° apart. Using this approximation, the magnitude of the total current can be computed by applying the Pythagorean theorem to the current phasors; that is,

$$I_T = \sqrt{I_{R1}^2 + I_{L1}^2}$$

   (a) Compare the total current measured in R_{S1} (Table 28–1) with the current found by applying the Pythagorean theorem to the current phasors.

   (b) What factors account for differences between the two currents?

2. How does the coil resistance measured in step 2 affect the angle between the current in the resistor and the current in the inductor?

3. In Experiment 26, a 1.0 kΩ resistor was used as a current-sensing resistor. Why would this value be unsatisfactory in this experiment?

4. If the inductor were open, what would happen to each of the following?
   (a) the total current in the circuit ________________
   (b) the phase angle between the generator voltage and current ________________
   (c) the generator voltage ________________

5. If the frequency were increased, what would happen to each of the following?
   (a) the total current in the circuit ________________
   (b) the phase angle between the generator voltage and current ________________
   (c) the generator voltage ________________

6. How would the phase angle between the generator voltage and current have been affected in each case?
   (a) if the resistor (R_1) were larger ________________
   (b) if the inductor were larger ________________

TECH TIP REVIEW:
Each of the following questions relate to the TECH TIP circuit shown in Figure 17–47 of the Floyd text.

1. Assume that either the inductor or the resistor was known to be open in the *RL* filter module. Explain how you would know which component was open from ohmmeter measurements alone.

2. How would the open component from the preceding question affect the output in each case?

29 Series Resonance

OBJECTIVES:

After performing this experiment, you will be able to:

1. Compute the resonant frequency, Q, and bandwidth of a series resonant circuit.
2. Measure the parameters listed in objective 1.
3. Explain the factors affecting the selectivity of a series resonant circuit.

READING:

Floyd, *Principles of Electric Circuits,* Sections 18–1 through 18–3

MATERIALS NEEDED:

One 100 mH inductor
One 0.01 μF capacitor
One 100 Ω resistor
One 47 Ω resistor
For Further Investigation: 270 Ω resistor
Application Problem: Load resistor and capacitor to be determined by student

SUMMARY OF THEORY:

The reactance of inductors increases with frequency according to the equation

$$X_L = 2\pi f L$$

On the other hand, the reactance of capacitors decreases with frequency according to the equation

$$X_C = \frac{1}{2\pi f C}$$

Consider the series LC circuit shown in Figure 29–1(a). In any LC circuit, there is a frequency at which the inductive reactance is equal to the capacitive reactance. The point at which there is equal and opposite reactance is called *resonance*. By setting $X_L = X_C$, substituting the relations just given, and solving for f, it is easy to show that the resonant frequency of an LC circuit is

$$f_r = \frac{1}{2\pi\sqrt{LC}}$$

where f_r is the resonant frequency. Recall that reactance phasors for inductors and capacitors are drawn in opposite directions because of the opposite phase shift which occurs between inductors and capacitors. At series resonance these two phasors are added and cancel each other. This is illustrated in Figure 29–1(b). The current in the circuit is limited only by the total

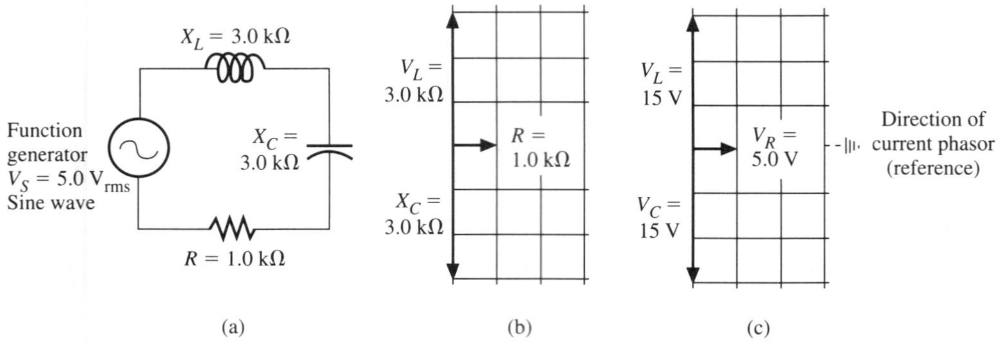

Figure 29–1

resistance of the circuit. The current in this example is 5.0 mA. If each of the impedance phasors is multiplied by this current, the result is the voltage phasor diagram shown in Figure 29–1(c). Notice that the voltage across the inductor and the capacitor can be *greater* than the applied voltage!

At the resonant frequency, the cancellation of the inductive and capacitive phasors leaves only the resistive phasor to limit the current in the circuit. Therefore, at resonance, the impedance of the circuit is a *minimum* and the current is a *maximum* and equal to V_S/R. The phase angle between the source voltage and current is zero. If the frequency is lowered, the inductive reactance will be smaller and the capacitive reactance will be larger. The circuit is said to be capacitive because the source current leads the source voltage. If the frequency is raised above resonance, the inductive reactance increases and the capacitive reactance decreases. The circuit is said to be inductive.

The *selectivity* of a resonant circuit describes how the circuit responds to a group of frequencies. A highly selective circuit responds to a narrow group of frequencies and rejects other frequencies. The *bandwidth* of a resonant circuit is the frequency range at which the current is 70.7% of the maximum current. A highly selective circuit thus has a narrow bandwidth. The sharpness of the response to frequencies is determined by the circuit Q. The Q for a series resonant circuit is the reactive power in either the coil or capacitor divided by the true power which is dissipated in the total resistance of the circuit. The bandwidth and resonant frequency can be shown to be related to the circuit Q by the equation

$$Q = \frac{f_r}{BW}$$

Figure 29–2 illustrates how the bandwidth can change with Q. Responses 1 and 2 have the same resonant frequency but different bandwidths. The bandwidth for curve 1 is shown. Response curve 2 has a higher Q and a smaller BW. A useful equation that relates the circuit resistance, capacitance, and inductance to Q is

$$Q = \frac{1}{R}\sqrt{\frac{L}{C}}$$

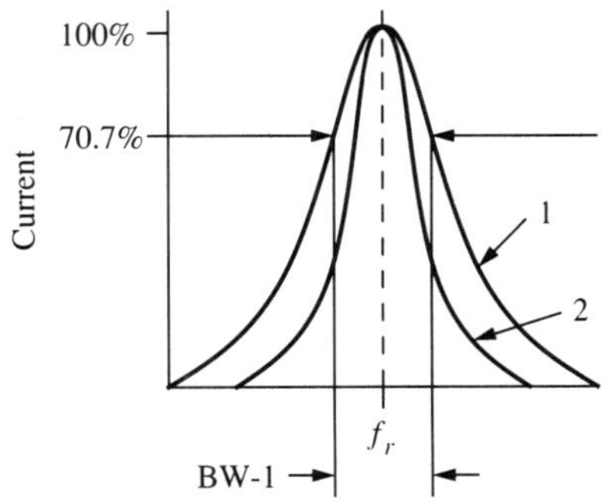

Figure 29–2

The value of R in this equation is the total equivalent series resistance in the circuit. Using this equation, the circuit response can be tailored to the application. For a highly selective circuit, the circuit resistance is held to a minimum, and the L/C ratio is made high.

The Q of a resonant circuit can also be computed from the equation

$$Q = \frac{X_L}{R}$$

where X_L is the inductive reactance and R is again the total equivalent series resistance of the circuit. The result is the same if X_C is used in the equation, since the values are the same at resonance, but X_L is usually shown because the resistance of the inductor is frequently the dominant resistance of the circuit.

PROCEDURE:

1. Measure the value of a 100 mH inductor, a 0.01 μF capacitor, a 100 Ω resistor, and a 47 Ω resistor. Enter the measured values in Table 29–1. If you cannot measure the inductor or capacitor, use the listed values.

2. Measure the resistance of the inductor. Enter the measured inductor resistance in Table 29–1.

3. Construct the circuit shown in Figure 29–3. The purpose of the parallel 100 Ω resistor is to reduce the Thevenin driving impedance of the function generator and, therefore, the total equivalent series resistance of the circuit.* Compute the total resistance, R_T, of the equivalent series circuit. Note that looking back to the generator, R_{TH} is in parallel with R_1. In equation form, the equivalent series resistance, R_T, is:

$$R_T = (R_{TH} \parallel R_1) + R_{COIL} + R_{S1}$$

Enter the computed total resistance in Table 29–2.

---

*If you have a low-impedance generator (50 Ω), R_1 is not necessary. To check your generator impedance, see Experiment 12, For Further Investigation.

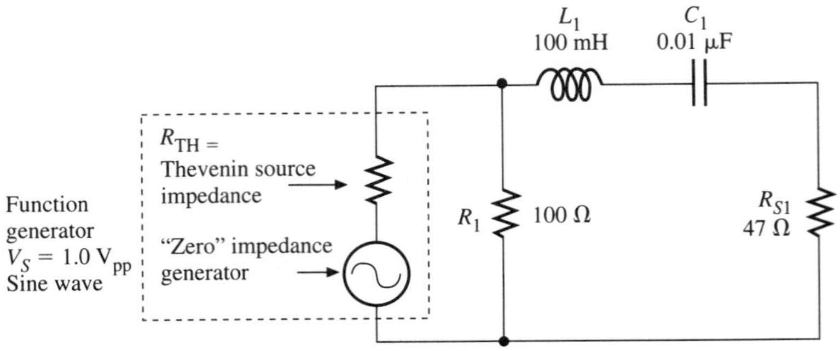

Figure 29–3

4. Using the measured values from Table 29–1, compute the resonant frequency of the circuit from the equation

$$f_r = \frac{1}{2\pi\sqrt{LC}}$$

Record the computed resonant frequency in Table 29–2.

5. Use the total resistance computed in step 3 and the measured values of L and C to compute the approximate Q of the circuit from the equation

$$Q = \frac{1}{R_T}\sqrt{\frac{L}{C}}$$

Enter the computed Q in Table 29–2.

6. Compute the bandwidth from the equation

$$BW = \frac{f_r}{Q}$$

Enter this as the computed BW in Table 29–2.

7. Using your oscilloscope, tune for resonance by observing the voltage across the sense resistor, R_{S1}. As explained in Floyd's text, the current in the circuit rises to a maximum at resonance. The sense resistor will have the highest voltage across it at resonance. Measure the resonant frequency with the oscilloscope. Record the measured resonant frequency in Table 29–2.

8. Check that the voltage across R_1 is 1.0 V_{pp}. Measure the peak-to-peak voltage across the sense resistor at resonance. The voltage across R_{S1} is directly proportional to the current in the series LC branch, so it is not necessary to compute the current. Record in Table 29–2 the measured peak-to-peak voltage across R_{S1} (V_{RS1}).

9. Raise the frequency of the generator until the voltage across R_{S1} falls to 70.7% of the value read in step 7. Do not readjust the generator's amplitude in this step. Measure and record this frequency as f_2 in Table 29–2.

10. Lower the frequency to below resonance until the voltage across R_{S1} falls to 70.7% of the value read in step 6. Again, do not adjust the generator amplitude. Measure and record this frequency as f_1 in Table 29–2.

11. Compute the bandwidth by subtracting f_1 from f_2. Enter this as the measured bandwidth in Table 29–2.

12. At resonance, the current in the circuit, the voltage across the capacitor, and the voltage across the inductor are all at a maximum value. Tune across resonance by observing the voltage across the capacitor; then try it on the inductor. Use the oscilloscope difference function technique described in Experiment 18. What is the maximum voltage observed on the capacitor? Is it the same or different than the maximum voltage across the inductor? Describe your observations in the conclusion section of your report.

FOR FURTHER INVESTIGATION:
In this experiment, you measured three points on the response curve similar to Figure 29–2. Using the technique of measuring the voltage across R_{S1}, find several more points on the response curve. Then change R_{S1} to 270 Ω and measure the new response. Graph both curves on Plot 29–1.

APPLICATION PROBLEM:
Series resonant circuits find application in certain matching networks. Consider the circuit shown in Figure 29–4(a). The Thevenin source is not always a resistive source but in many cases can have a component of inductance or capacitance. The Thevenin impedance and voltage are not adjustable. Assume you need to find the impedance of the load that will absorb the maximum power from the source at some specific frequency. If the Thevenin source contains an inductive component, $X_{L\text{-}TH}$, then the load impedance can be made capacitive by an equal amount ($X_{C\text{-}L} = X_{C\text{-}TH}$) to cancel the reactive effect. Likewise, if the Thevenin source contains a capacitive component, then the load impedance can be made inductive. By adding the appropriate reactive component to the load, a series resonance circuit is constructed that

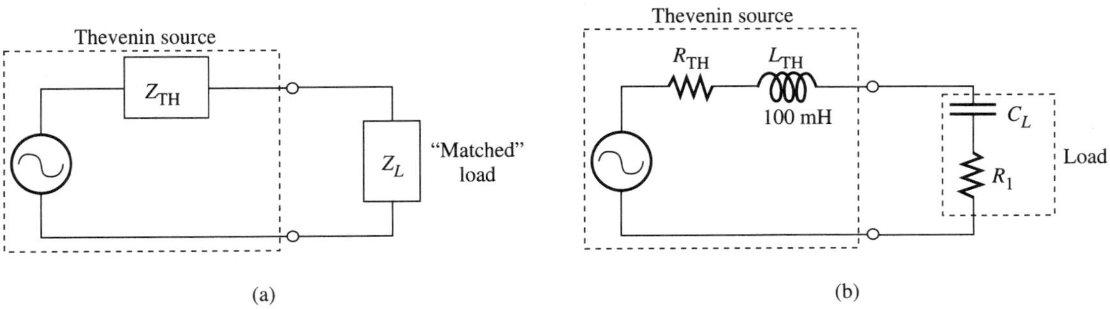

Figure 29–4

cancels the effect of the reactive portion of the Thevenin source. The maximum power is then transferred if, in addition to the reactance component, the load contains a resistance component equal to the Thevenin resistance of the source.

For this problem, assume the Thevenin source is your function generator in series with the inductance used in this experiment. (This type of driving impedance occurs with transformer coupled amplifiers.) The operating frequency is 50 kHz. Construct the circuit shown in Figure 29–4(b). Set the generator for 1.0 V_{pp} at 50 kHz. Choose an appropriate value of capacitance to resonate with the inductive component of the source and a resistance to match the source resistance. (Don't forget that the source resistance includes the coil resistance.) Prove that your load absorbs the maximum power from the load by calculating the load power from voltage measurements above and below the operating frequency.

Report for Experiment 29

Name ________________
Date ________________
Class ________________

OBJECTIVES:

PROCEDURE:

DATA:

Table 29–1

	Listed Value	Measured Value
L_1	100 mH	
C_1	0.01 μF	
R_1	100 Ω	
R_{S1}	47 Ω	
L_1 resistance		

Table 29–2

	Computed	Measured
R_T		
f_r		
Q		
V_{RS1}		
f_2		
f_1		
BW		

RESULTS AND CONCLUSION:

FURTHER INVESTIGATION RESULTS:

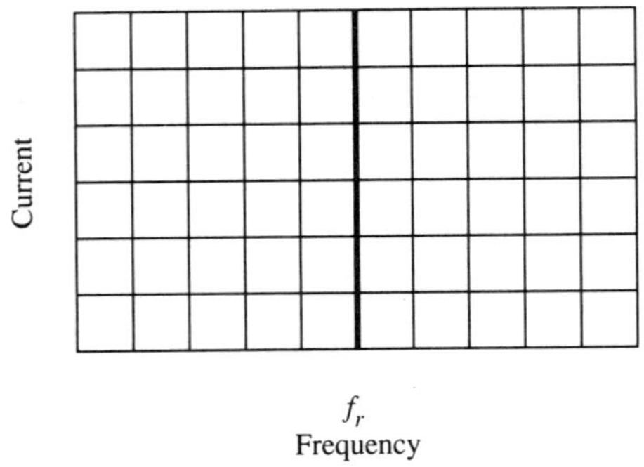

Plot 29–1

APPLICATION PROBLEM RESULTS:

EVALUATION AND REVIEW QUESTIONS:

1. (a) Determine the percent difference between the computed and measured bandwidth.

   (b) What factors account for the difference between the computed and measured values?

2. (a) What is the total impedance of the experimental circuit at resonance?

   (b) What is the phase shift between the total current and voltage?

3. (a) In step 12, you measured the maximum voltage across the capacitor and the inductor. The maximum voltage across either one should have been larger than the source voltage. How do you account for this?

   (b) Is this a valid technique for finding the resonant frequency?

4. (a) **What would happen to the resonant frequency if the inductor were twice as large and the capacitor were half as large?**

   (b) **What would happen to the bandwidth?**

5. (a) Compute the resonant frequency for a circuit consisting of a 50 μH inductor in series with a 1000 pF capacitor.

(b) If the total resistance of the above circuit is 10 Ω, what are the Q and the bandwidth?

6. In a series resonant circuit, does the circuit look capacitive or inductive below the resonant frequency? Explain your answer.

30 Parallel Resonance

PSpice

OBJECTIVES:

After performing this experiment, you will be able to:

1. Compute the resonant frequency, Q, and bandwidth of a parallel resonant circuit.
2. Measure the frequency response of a parallel resonant circuit.
3. Use the frequency response curve to determine the bandwidth of a parallel resonant circuit.

READING:

Floyd, *Principles of Electric Circuits,* Sections 18–6 through 18–9

MATERIALS NEEDED:

One 100 mH inductor
One 0.047 μF capacitor
One 1.0 kΩ resistor
For Further Investigation: Second 0.01 μF capacitor, one 1000 pF capacitor
Application Problem: Sweep generator for audio frequencies (to 10 kHz).

SUMMARY OF THEORY:

In an *RLC* parallel circuit, the current in each branch is determined by the applied voltage and the impedance of that branch. For an "ideal" inductor (no resistance), the branch impedance is X_L, and for a capacitor the branch impedance is X_C. Since X_L and X_C are functions of frequency, it is apparent that the currents in each branch are also dependent on the frequency. For any given L and C, there is a frequency at which the currents in each are equal and of opposite phase. This frequency is the resonant frequency and is found using the same equation as was used for series resonance:

$$f_r = \frac{1}{2\pi\sqrt{LC}}$$

The circuit and phasor diagram for an ideal parallel *RLC* circuit at resonance are illustrated in Figure 30–1. Some interesting points to observe are: The total source current at resonance is equal to the current in the resistor. The total current is actually *less* than the current in either the inductor or the capacitor. This is because of the opposite phase shift that occurs between inductors and capacitors, causing the addition of the currents to cancel. Also, in the ideal case, the impedance of the circuit is solely determined by R, as the inductor and capacitor appear to be open at resonance. In a two-branch circuit consisting of only L and C, the source current would be zero, causing the impedence to be infinite! Of course, this does not happen with actual components, which do have resistance and other effects.

In a practical two-branch parallel circuit consisting of an inductor and a capacitor, the only significant resistance is the winding resistance of the inductor. Figure 30–2(a) illustrates a practical parallel *LC* circuit containing winding resistance. By network theorems, the practical *LC* circuit can be converted to an equivalent parallel *RLC* circuit, as shown in Figure 30–2(b).

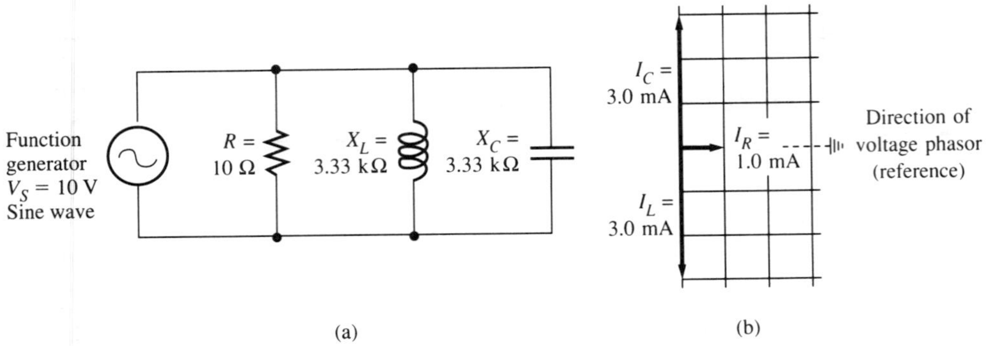

Figure 30–1

The equivalent circuit is easier to analyze. The phasor diagram for the ideal parallel *RLC* circuit can then be applied to the equivalent circuit, as was illustrated in Figure 30–1. The equations to convert the inductance and its winding resistance to an equivalent parallel circuit are

$$L_{\text{eq}} = L\left(\frac{Q^2 + 1}{Q^2}\right) \qquad R_{\text{peq}} = R_W(Q^2 + 1)$$

where R_{peq} represents the parallel equivalent resistance, Q represents the Q of the inductor, and R_W represents the winding resistance of the inductor.

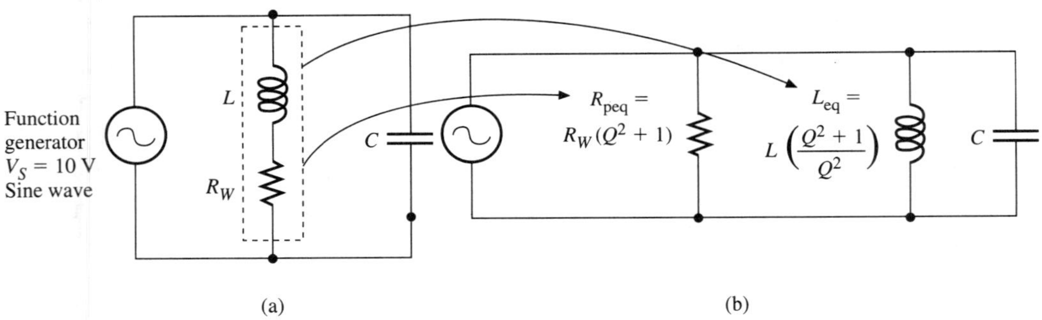

Figure 30–2

The Q used in the conversion equation is the Q for the inductor:

$$Q = \frac{X_L}{R_W}$$

The *selectivity* of series resonant circuits was discussed in Experiment 26. Parallel resonant circuits also respond to a group of frequencies. In parallel resonant circuits, the impedance as a function of frequency has the same shape as the current versus frequency curve for series resonant circuits. The *bandwidth* of a parallel resonant circuit is the frequency range at which the circuit impedance is 70.7% of the maximum impedance. The sharpness of the response to frequencies is again measured by the circuit Q. The circuit Q will be different than the Q of the inductor if there is additional resistance in the circuit. If there is no additional resistance in parallel with L and C, then the Q for a parallel resonant circuit is equal to the Q of the inductor.

PROCEDURE:

1. Measure the value of a 100 mH inductor, a 0.047 μF capacitor, and a 1.0 kΩ resistor. Enter the measured values in Table 30–1. If it is not possible to measure the inductor or capacitor, use the listed values.

2. Measure the resistance of the inductor. Enter the measured inductor resistance in Table 30–1.

3. Construct the circuit shown in Figure 30–3. The purpose of R_{S1} is to develop a voltage that can be used to sense the total current in the circuit. Compute the resonant frequency of the circuit using the equation

$$f_r = \frac{1}{2\pi\sqrt{LC}}$$

Enter the computed resonant frequency in Table 30–2. Set the generator to the f_r at 1.0 V_{pp} output as measured with your oscilloscope. Use peak-to-peak values for all voltage measurements in this experiment.

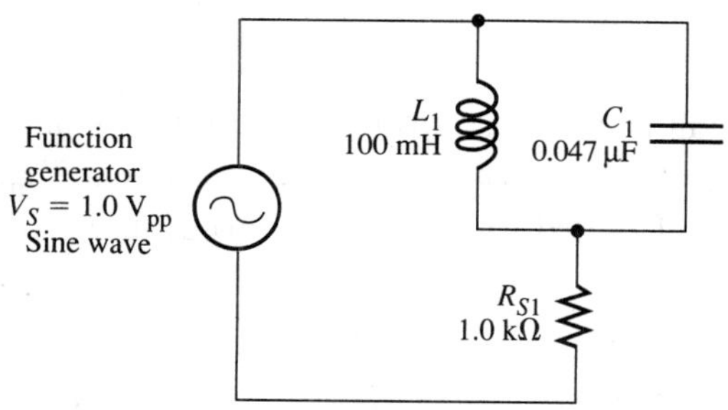

Figure 30–3

4. The Q of a parallel LC circuit with no resistance other than the inductor winding resistance is equal to the Q of the inductor. Compute the approximate Q of the parallel LC circuit from

$$Q = \frac{X_L}{R_W}$$

Enter the computed Q in Table 30–2.

5. Compute the bandwidth from the equation

$$BW = \frac{f_r}{Q}$$

Enter this as the computed *BW* in Table 30–2.

6. Connect your oscilloscope across R_{S1} and tune for resonance by observing the voltage across the sense resistor, R_{S1}. Resonance occurs when the voltage across R_{S1} is a minimum, since the impedance of the parallel *LC* circuit is highest. Measure the resonant frequency (f_r) and record the measured result in Table 30–2.

7. Compute a frequency increment (f_i) by dividing the measured bandwidth by 4; that is,

$$f_i = \frac{BW}{4}$$

Enter the computed f_i in Table 30–2.

8. Use the measured resonant frequency (f_r) and the frequency increment (f_i) from Table 30–2 to compute 11 frequencies according to the *Computed Frequency* column of Table 30–3. Enter the 11 frequencies in column 1 of Table 30–3.

9. Tune the generator to each of the computed frequencies listed in Table 30–3. At each frequency, check that the generator voltage is still at 1.0 V_{pp}; then measure the peak-to-peak voltage across R_{S1}. Record the voltage in column 2 of Table 30–3.

10. Compute the total peak-to-peak current, *I*, at each frequency by applying Ohm's law to the sense resistor R_{S1} (that is, $I = V_{RS1}/R_{S1}$). Record the current in column 3 of Table 30–3.

11. Use Ohm's law with the measured source voltage (1.0 V_{pp}) and source current at each frequency to compute the impedance at each frequency. Complete Table 30–3 by listing the computed impedances.

12. On Plot 30–1, draw the impedance-versus-frequency curve. From your curve determine the bandwidth. Complete Table 30–2 with the measured bandwidth.

FOR FURTHER INVESTIGATION:

The resonant circuit in this experiment can be made to "ring" by pulsing it every so often with a pulse generator. The analogy of a child swinging is appropriate—the swinging motion will continue (and eventually die out) unless it receives a "kick" every so often. The kick in this case will be provided by a square wave from the function generator, which will be capacitively coupled to the circuit as shown in Figure 30–4. Set the generator for 500 Hz. Observe the waveform across the parallel *LC* combination. (You may need to adjust TRIGGER HOLDOFF to obtain a stable trace.)

Try replacing C_A with a 1000 pF capacitor and observe the effect. (Can you explain it?) How does the frequency compare with that found in the experiment? Sketch the waveforms in your report.

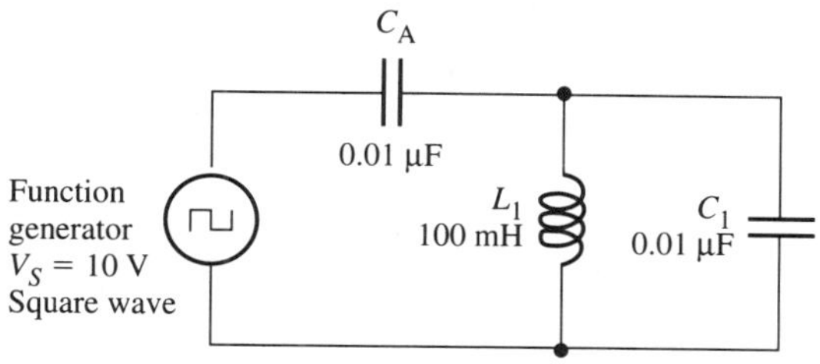

Figure 30–4

APPLICATION PROBLEM:

Parallel resonant circuits are used in many circuits, such as oscillators, where a specific frequency response is required. It is useful to be able to display the frequency response of circuits with an oscilloscope. The oscilloscope can be used to display the resonant dip in current by connecting a sweep generator to the circuit. This converts the time base on the oscilloscope to a frequency base. The sweep generator produces an FM (frequency modulated), signal which is connected in place of the signal generator. In addition, the sweep generator has a synchronous sweep output, which should be connected to the oscilloscope on the "X" channel input. The "Y" channel input is connected across the 47 Ω sense resistor. The oscilloscope is placed in the X-Y mode. A diagram of the setup is shown in Figure 30–5. Build the circuit shown, determine a method to calibrate the frequency base, and summarize your procedure in your report.

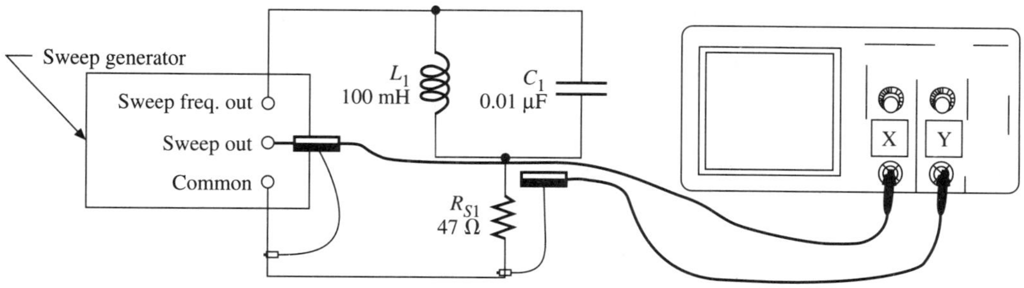

Figure 30–5

PSpice EXAMPLE:

The following PSpice example will plot the voltage in the sense resistor, R_{S1}, as frequency is incremented.

```
LAB 30 FIG 30-3
VS 1 0 AC 1V
L1 1 2 100E-3
C1 1 2 .047E-6
RS1 2 0 1E3
.AC LIN 50 500Hz 8000Hz
.PLOT AC V(RS1)
.PROBE
.OPTIONS NOPAGE
.END
```

Report for Experiment 30

Name ____________
Date ____________
Class ____________

OBJECTIVES:

PROCEDURE:

DATA:

Table 30–1

	Listed Value	Measured Value
L_1	100 mH	
C_1	0.05 μF	
R_{S1}	1.0 kΩ	
L_1 resistance		

Table 30–2

	Computed	Measured
f_r		
Q		
BW		
$f_i = \frac{BW}{4}$		

Table 30–3

Computed Frequency	V_{RS1}	I	Z
$f_r - 5f_i =$			
$f_r - 4f_i =$			
$f_r - 3f_i =$			
$f_r - 2f_i =$			
$f_r - 1f_i =$			
$f_r =$			
$f_r + 1f_i =$			
$f_r + 2f_i =$			
$f_r + 3f_i =$			
$f_r + 4f_i =$			
$f_r + 5f_i =$			

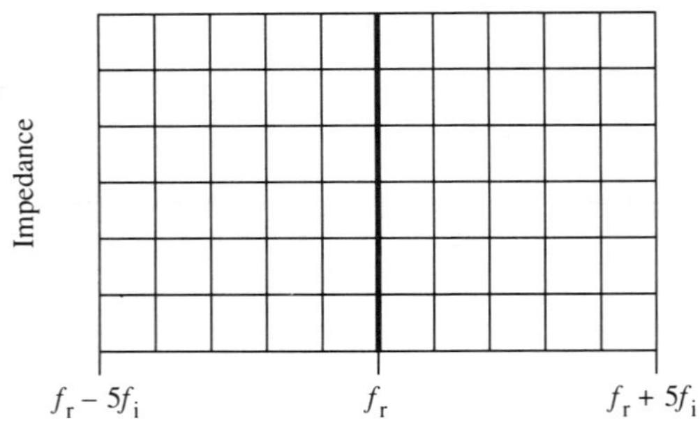

Plot 30–1

RESULTS AND CONCLUSION:

FURTHER INVESTIGATION RESULTS:

APPLICATION PROBLEM RESULTS:

EVALUATION AND REVIEW QUESTIONS:

1. (a) Compare the impedance as a function of frequency for series and parallel resonance.

   (b) Compare the current as a function of frequency for series and parallel resonance.

2. What was the phase shift between the source current and voltage at resonance?

3. At resonance the total current was a minimum but the branch currents were not. How could you find the value of the current in each branch?

4. What factors affect the Q of a parallel resonant circuit?

5. In the circuit of Figure 30–2a, assume the inductor is 100 mH with 120 Ω of winding resistance and the capacitor is 0.01 μF. Compute each of the following.

   (a) the resonant frequency ____________

   (b) the reactance, X_L, of the inductor at resonance ____________

   (c) the Q of the circuit ____________

   (d) the bandwidth, BW ____________

6. In a parallel resonant circuit, does the circuit look capacitive or inductive below the resonant frequency? Explain your answer.

TECH TIP REVIEW:
Each of the following questions relate to the TECH TIP circuit shown in Figure 18–55 of the Floyd text.

1. As you know, the inductive and capacitive reactances are each a function of the frequency of the resonant circuit. Compute the inductive reactance of the coil when the resonant circuit is tuned to a station at the lower and the upper end of the dial 535 kHz and 1605 kHz.

2. Assume the inductor has a winding resistance of 15 Ω. Compute the Q of the inductor at the lower and the upper end of the dial. Which end of the dial do you expect to have the greater selectivity?

31 Passive Filters

PSpice

OBJECTIVES:

After performing this experiment, you will be able to:

1. Compare the characteristics and responses of low-pass, high-pass, bandpass, and notch filters.
2. Construct a T filter, a pi filter, and a resonant filter circuit and measure their frequency responses.

READING:

Floyd, *Principles of Electric Circuits,* Sections 19–1 through 19–4 and 19–6

MATERIALS NEEDED:

Resistors:

One 680 Ω, one 1.6 kΩ

Capacitors:

One 0.033 μF, two 0.1 μF

One 100 mH inductor

Application Problem: One 4.7 kΩ resistor, two 10 kΩ resistors, one 2000 pF capacitor, two 1000 pF capacitors

SUMMARY OF THEORY:

In many circuits, different frequencies are present. If certain frequencies are not desired, they can be rejected with special circuits called *filters*. Filters can be designed to pass either low or high frequencies. For example, in communication circuits, an audio-frequency (AF) signal may be present with a radio-frequency (RF) signal. The AF signal could be retained and the RF signal rejected with a *low-pass* filter. A *high-pass* filter will do the opposite: It will pass the RF signal and reject the AF signal. Sometimes the frequencies of interest are between other frequencies that are not desired. This is the case for a radio or television receiver, for example. The desired frequencies are present along with many other frequencies coming into the receiver. A resonant circuit is used to select the desired frequencies from the band of frequencies present. A circuit that passes only selected frequencies from a band is called a *bandpass* filter. The opposite of a bandpass filter is a *band-reject,* or *notch,* filter. A typical application of a notch filter is to eliminate a specific interfering frequency from a band of desired frequencies. Figure 31–1 illustrates representative circuits and the frequency responses of various types of filters.

The simplest filters are the *RC* and *RL* series circuits studied in Experiments 25 and 27. These circuits can be used as either high-pass or low-pass filters, depending on where the input and output voltages are applied and removed. A problem with simple *RC* or *RL* filters is that they change gradually from the passband to the stop band. You illustrated this characteristic on Plot 25–1.

Improved filter characteristics can be obtained by combining several filter sections together. Unfortunately, you cannot simply stack identical sections together to improve the response, because there are loading effects that must be taken into account. Two common

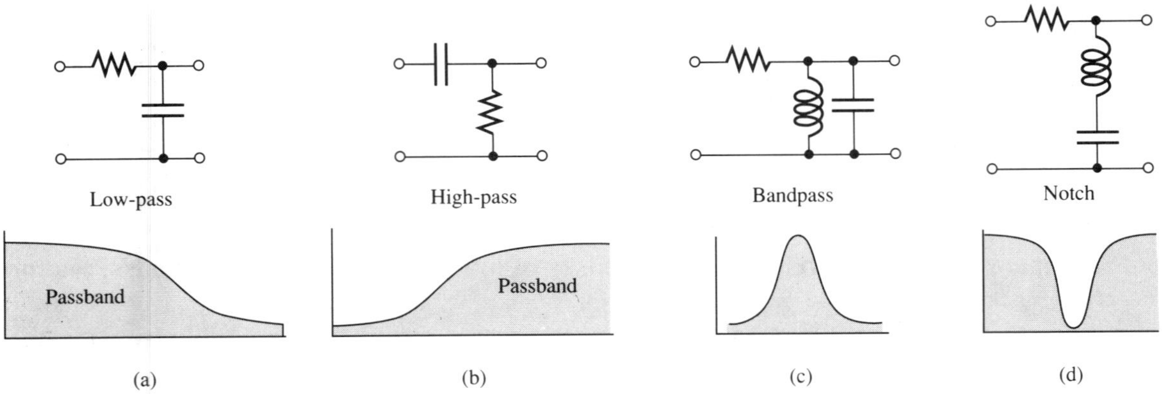

Figure 31–1

improved filters are the *T* and the *pi* filters, so named because of the placement of the components in the circuit. Examples of T and pi filters are shown in Figure 31–2. Notice that the low-pass filters have an inductor in series with the load and a capacitor in parallel with the load. The high-pass filter is the opposite.

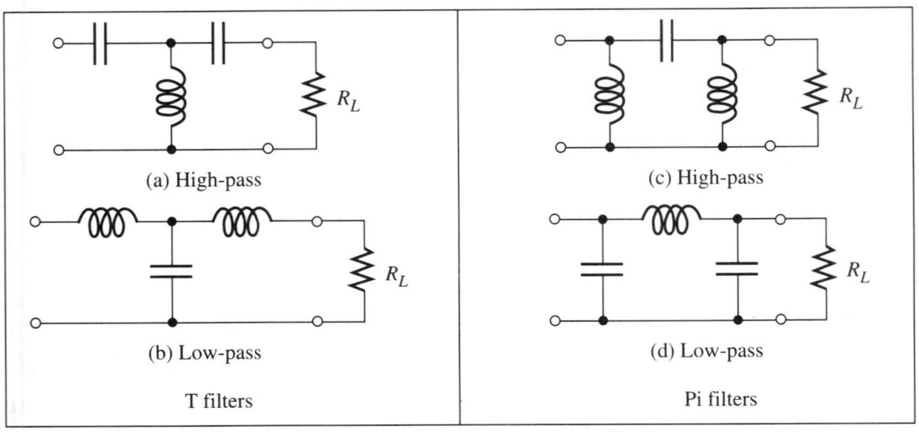

Figure 31–2

The choice of using a T or a pi filter is determined by the load resistor and source impedance. If the load resistor is much larger than the source impedance, then the T-type filter is best. If the load resistor is much lower than the source impedance, then the pi filter is best.

PROCEDURE:

1. Obtain the components listed in Table 31–1. For this experiment, it is important to have values that are close to the listed ones. Measure all components and record the measured values in Table 31–1. Use listed values for those components that you cannot measure.

2. Construct the pi filter circuit illustrated in Figure 31–3. Using the oscilloscope, set the function generator for a 500 Hz sine wave at 6.0 V_{pp}. The generator voltage should be measured with the circuit connected. Check both voltage and frequency with the oscilloscope. All voltages throughout this experiment are specified and recorded as peak-to-peak values.

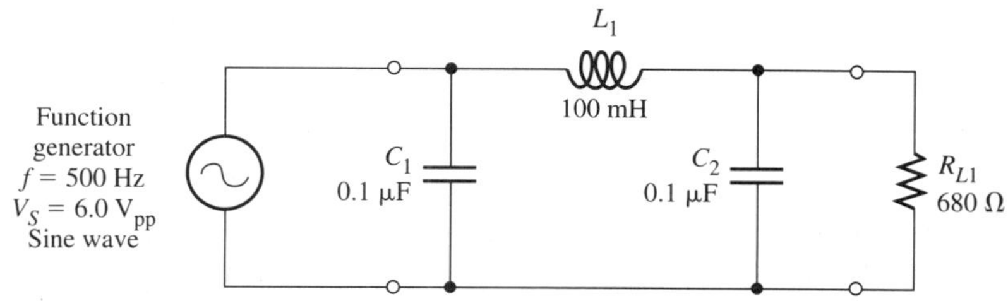

Figure 31–3

3. Measure the peak-to-peak voltage across the load resistor (V_{RL1}) at 500 Hz. Record the measured voltage in Table 31–2.

4. Change the frequency of the generator to 1000 Hz. Adjust the generator's amplitude to 6.0 V_{pp} and again check both voltage and frequency with the oscilloscope. Measure V_{RL1}, entering the data in Table 31–2. Continue in this manner for each frequency listed in Table 31–2.

5. Graph the voltage across the load resistor (V_{RL1}) as a function of frequency on Plot 31–1.

6. Construct the T filter circuit illustrated in Figure 31–4. Set the signal generator for a 500 Hz sine wave at 6.0 V_{pp}. The generator voltage should be measured with the circuit connected. Check both voltage and frequency with the oscilloscope, as before.

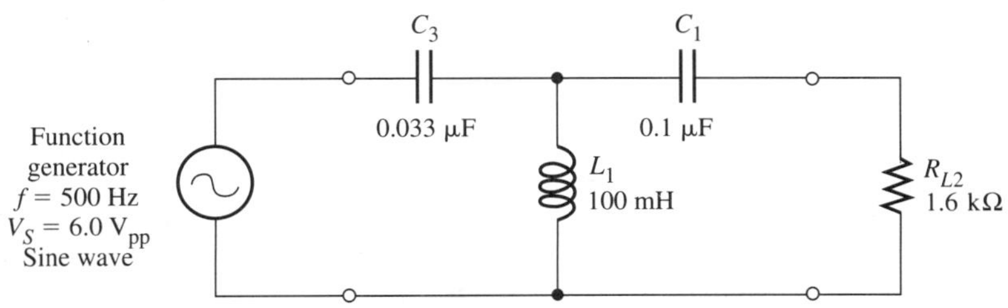

Figure 31–4

7. Measure and record the peak-to-peak voltage across the load resistor (V_{RL2}) for each frequency listed in Table 31–3. Keep the generator voltage at 6.0 V_{pp}. Graph the peak-to-peak voltage across the load resistor (V_{RL2}) as a function of frequency on Plot 31–2.

8. Construct the series resonant filter circuit illustrated in Figure 31–5. Set the generator for 6.0 V_{pp} at 500 Hz.

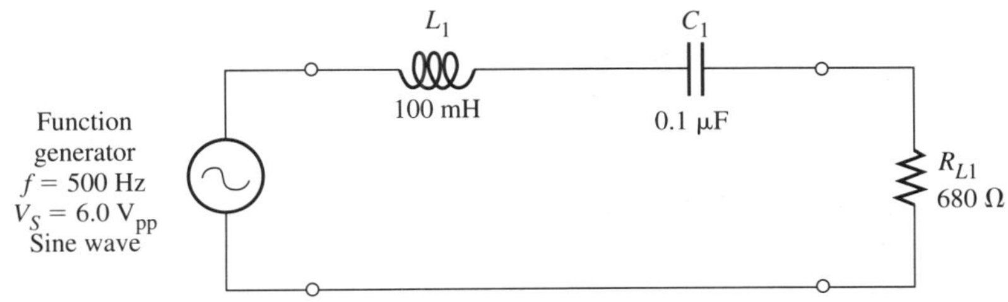

Figure 31–5

9. Measure and record the peak-to-peak voltage across the load resistor (V_{RL1}) for each frequency listed in Table 31–4. Graph the voltage across the load resistor as a function of frequency on Plot 31–3.

FOR FURTHER INVESTIGATION:

Using the components from this experiment, construct a parallel resonant notch (band-reject) filter. Measure the frequency response with a sufficient number of points to determine the bandwidth accurately. The bandwidth *(BW)* of a resonant filter is the difference in the two frequencies at which the response is 70.7% of the maximum output. From your data, determine the *BW* of the parallel notch resonant filter.

Summarize your results, including the notch frequency, *BW,* and response curve in your report.

APPLICATION PROBLEM:

An interesting application of a notch filter is the twin-T oscillator shown in Figure 31–6(a). It oscillates at the notch frequency given by the equation

$$f_r = \frac{1}{2\pi RC}$$

Test the notch filter portion of the oscillator by constructing the filter shown in Figure 31–6(b). Connect a function generator to the input and set it for a sine wave at 6.0 V_{pp}. Compare the computed and measured notch frequency for the filter. Plot the voltage out as a function of the frequency for several points at and near the notch frequency.

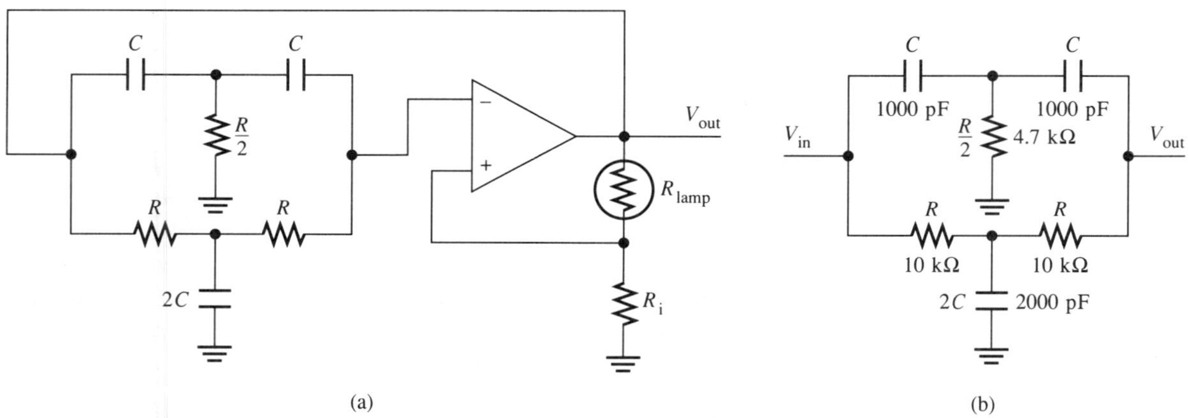

Figure 31–6

PSpice EXAMPLE:

The following PSpice example will plot the output voltage as a function of frequency for the filter shown in Figure 31–3.

```
LAB31 FIG 31–3
VS 1 0 AC 6V
C1 1 0 .1E–6
L1 1 2 100E–3
C2 2 0 .1E–6
RL1 2 0 680
.AC LIN 50 500Hz 8000Hz
.PLOT AC V(RL1)
.PROBE
.OPTIONS NOPAGE
.END
```

Report for Experiment 31

Name ______________________
Date ______________________
Class ______________________

OBJECTIVES:

PROCEDURE:

DATA:

Table 31–1

	Listed Value	Measured Value
L_1	100 mH	
C_1	0.1 μF	
C_2	0.1 μF	
C_3	0.033 μF	
R_{L1}	680 Ω	
R_{L2}	1.6 kΩ	

Table 31–2

Frequency	V_{RL1}
500 Hz	
1000 Hz	
1500 Hz	
2000 Hz	
3000 Hz	
4000 Hz	
8000 Hz	

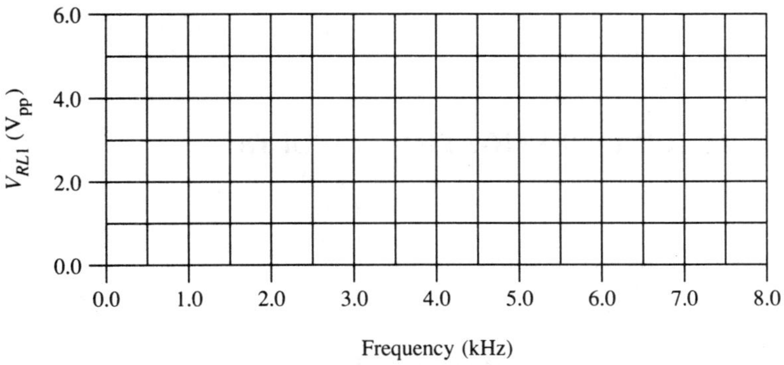

Plot 31–1

Table 31–3

Frequency	V_{RL2}
500 Hz	
1000 Hz	
1500 Hz	
2000 Hz	
3000 Hz	
4000 Hz	
8000 Hz	

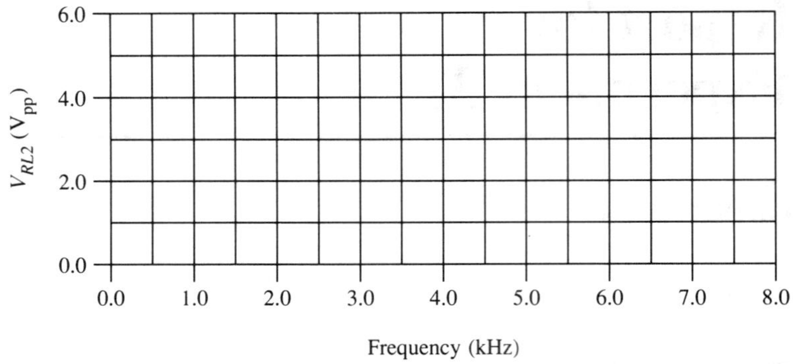

Plot 31–2

Table 31–4

Frequency	V_{RL1}
500 Hz	
1000 Hz	
1500 Hz	
2000 Hz	
3000 Hz	
4000 Hz	
8000 Hz	

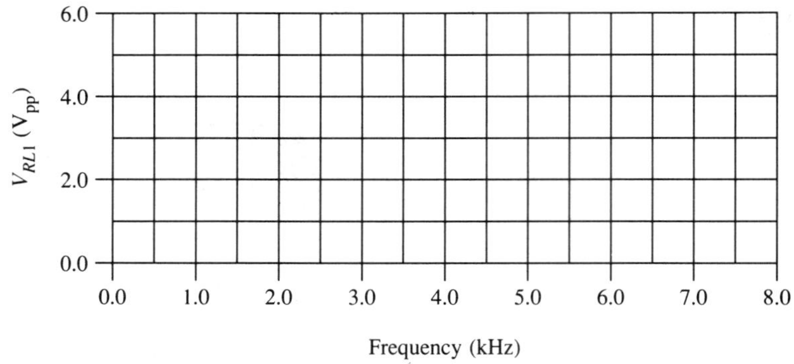

Plot 31–3

RESULTS AND CONCLUSION:

FURTHER INVESTIGATION RESULTS:

APPLICATION PROBLEM RESULTS:

EVALUATION AND REVIEW QUESTIONS:

1. The cutoff frequency for each filter in this experiment is that frequency at which the output is 70.7% of its maximum value. From the frequency response curves in Plots 31–1 and 31–2, estimate the cutoff frequency for the high- and low-pass filters.
   (a) Pi filter cutoff frequency ____________
   (b) T filter cutoff frequency ____________

2. Compare the response curve of the high and low filters in this experiment with the response curve from the simple *RC* filter in Experiment 22.

3. For each filter constructed in this experiment, identify it as a low-pass, high-pass, bandpass, or notch filter:
   (a) Plot 31–1 (pi filter) ____________
   (b) Plot 31–2 (T filter) ____________
   (c) Plot 31–3 (resonant filter) ____________

4. Explain what happens to the response curve from the series resonant filter if the output were taken across the inductor and capacitor instead of the load resistor.

5. Assume that the inductor in the series resonant filter in Figure 31–5 is shorted. What frequency response would you expect as a result?

6. With circuit sketches, show the difference between a parallel resonant circuit used as a bandpass filter and as a notch filter.

32 Circuit Theorems for AC Circuits

OBJECTIVES:
After performing this experiment, you will be able to:
1. Compute the Thevenin equivalent circuit for a complex impedance.
2. Test reactive loads on a circuit and its Thevenin equivalent.

READING:
Floyd, *Principles of Electric Circuits,* Sections 20–1 through 20–6

MATERIALS NEEDED:
Resistors:
One 3.3 kΩ, one 4.7 kΩ, one 9.1 kΩ, one 10 kΩ
Capacitors:
One 1000 pF, one 4700 pF, one 0.01 μF, one 0.1 μF
One 100 mH inductor

SUMMARY OF THEORY:
All of the important theorems from dc circuits can be applied to ac circuits by employing phasor mathematics. This includes Kirchhoff's laws, the superposition theorem, Thevenin's theorem, Norton's theorem, and so forth. There are no new theorems to be introduced for ac circuits that have not already been presented. Even ideas such as loop and node equations can be set up provided you use phasors. If you are uncertain about any theorems, you should review the appropriate material in the dc sections.

As an example of the application of circuit theorems to ac circuits, consider the circuit shown in Figure 32–1. This circuit is similar to many amplifier circuits, with both high and low cutoff frequencies due to parallel and series capacitances. Thevenin's theorem can be applied to this circuit at a specified frequency. The Thevenin voltage is defined in the same manner as in a dc circuit—it is the voltage that appears at the terminals of the circuit (**A-B**) with no load. The Thevenin impedance is the impedance seen looking from the output terminals (**A-B**) with the source replaced with its internal impedance. These definitions require the computation of phasor quantities and are dependent on the frequency of the generator. Other than this, they are equivalent to the dc definitions.

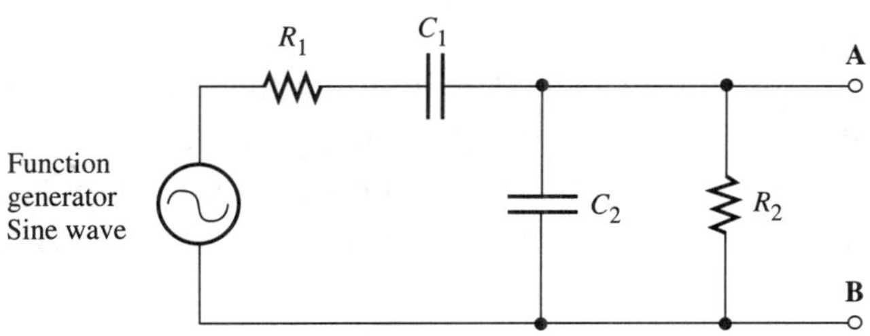

(a) Circuit

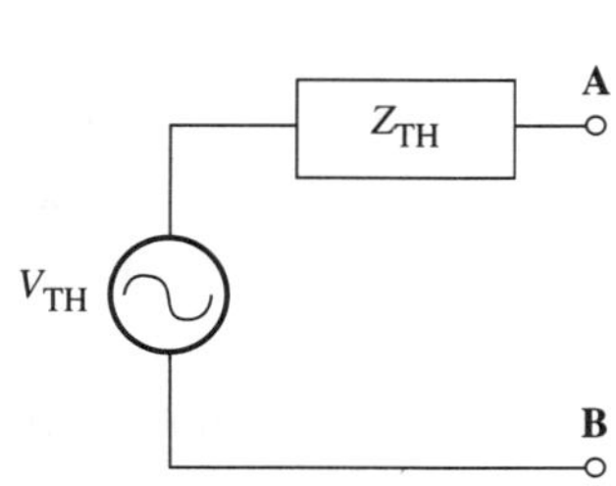

(b) Thevenin equivalent

Figure 32–1

At very low frequencies, C_1 appears to be open causing the Thevenin voltage to be near zero and the Thevenin impedance to be primarily determined by the parallel resistance, R_2. At very high frequencies, the reactance of C_2 is small, which causes both the Thevenin voltage and impedance also to appear small.

PROCEDURE:

1. Measure the components listed in Table 32–1 and record the measured values. For this experiment, try to obtain values that are within 10% or better of the listed values.

2. Construct the circuit shown in Figure 32–2(a). Set the function generator for a 3.0 V_{pp} sine wave at 17 kHz while it is connected to the circuit. Carefully check the voltage and frequency with your oscilloscope.

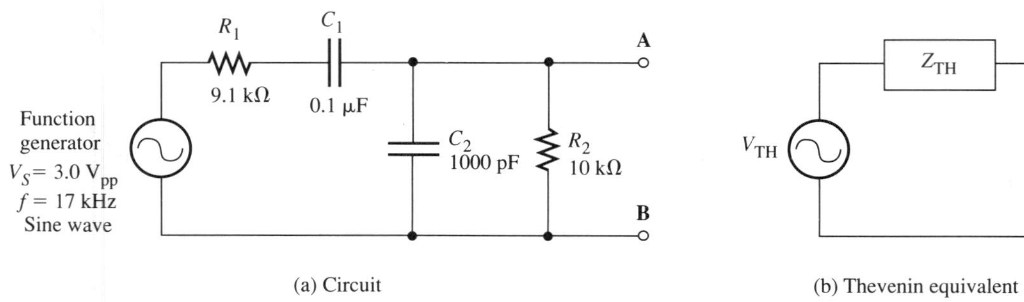

Figure 32–2

3. In this step, you will test a capacitive load, an inductive load, and a resistive load on your circuit. Place the loads, one at a time, across the **A-B** terminals of the circuit. Using your oscilloscope, measure the peak-to-peak output voltage and the phase angle with respect to the generator. (See Experiment 27 if you need to review phase angle measurements.) Enter your measured values in Table 32–2.

4. Compute the Thevenin impedance of the circuit. The reactance of C_1 is small at the operating frequency, so it can be ignored in the calculation. You need to include the impedance of the signal generator in your calculation. Looking from the output terminals, the generator impedance is in series with R_1, and together they are in parallel with C_2 and R_2. Show your calculation in the space provided in the report. Your result should indicate that the Thevenin impedance is nearly equivalent to a 3.9 kΩ resistance in series with a 4700 pF capacitor.

5. Compute the Thevenin voltage for the circuit. The Thevenin voltage can be computed by first finding the impedance of the parallel combination of C_2 and R_2. The generator sees this complex impedance in series with R_1 and its own internal impedance, so you can apply the voltage divider rule to obtain the unloaded output voltage. Show your calculation in your report.

6. Construct the circuit shown in Figure 32–2(b). Set the generator for the Thevenin voltage you found in step 5. The total Thevenin resistance is approximately 3.9 kΩ. You need to include the generator impedance in this value (i.e., if you are using a 600 Ω

generator, you will need to use a 3.3 kΩ resistor). This resistance is in series with a 4700 pF capacitor as computed in step 4.

7. Test each of the loads that you tested in step 3 on the Thevenin circuit. Measure both the peak-to-peak voltage and the phase angle as before. Tabulate your results in Table 32–2.

FOR FURTHER INVESTIGATION:

Kirchhoff's voltage law can be applied to an ac circuit. Use Kirchhoff's voltage law to compute the voltage drop across R_1 in the circuit of Figure 32–2(a). Then measure the magnitude of the voltage across R_1 using the difference function technique described in Experiment 18. Compare your calculated and measured values in your report.

APPLICATION PROBLEM:

A notch filter in a twin-T oscillator was shown in the application problem for Experiment 31. It is possible to find the voltage drops throughout the circuit and V_{out} by writing loop equations but it is necessary to use phasor math. The circuit is redrawn in Figure 32–3 with three loops shown (and substituting $R/2 = 5.0\ \text{k}\Omega$). Assume the generator is set for 1.0 V and is tuned to the notch frequency. At this frequency, $R = -jX_C = 10\ \text{k}\Omega$. (Note that the reactance of the 2000 pF capacitor is $-j\ (5\ \text{k}\Omega)$). The loop currents can be found using determinants with phasor math. If you do this, you will find that at the notch frequency for $V_{in} = 1.0$ V:

$$I_A = -0.075 - j0.025 \text{ mA}$$
$$I_B = -0.025 + j0.025 \text{ mA}$$
$$I_C = -0.025 + j0.075 \text{ mA}$$

These loop currents can now be used to compute the voltage drops across each component in the circuit. Compute the voltage drop across each component, and list the computed voltage drops in your report. V_{out} can be computed by summing the voltages from ground to the output point (as vectors) across R and $2C$ or by adding the voltage across $R/2$ and C. Calculate V_{out} and show that it is equal to zero at the notch frequency.

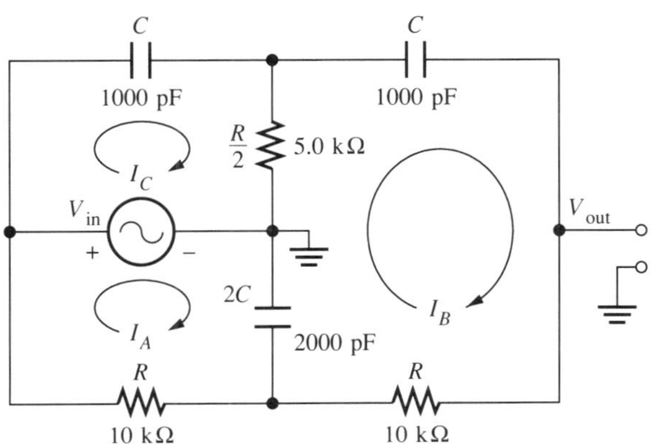

Figure 32–3

Report for Experiment 32

Name ______________
Date ______________
Class ______________

OBJECTIVES:

PROCEDURE:

DATA:

Table 30–1

	Listed Value	Measured Value
C_1	0.1 μF	
C_2	1000 pF	
R_1	9.1 kΩ	
R_2	10 kΩ	

Table 32–2

		Original Circuit		Thevenin Circuit	
Load	Listed Value	Measured Voltage	Measured Phase Angle	Measured Voltage	Measured Phase Angle
C_L	0.01 μF				
L_L	100 mH				
R_L	4.7 kΩ				

Step 4 (Calculations):

Step 5 (Calculations):

RESULTS AND CONCLUSION:

FURTHER INVESTIGATION RESULTS:

APPLICATION PROBLEM RESULTS:

EVALUATION AND REVIEW QUESTIONS:

1. (a) Calculate the Thevenin voltage for the circuit in Figure 32–2 at a frequency of 100 Hz. What component can you safely ignore in this calculation?

   (b) Calculate the Thevenin voltage for the circuit in Figure 32–2 at a frequency of 100 kHz. What component can you safely ignore in this calculation?

2. If you wanted to deliver maximum power from the circuit in Figure 32-2, the load impedance should be the complex conjugate of the Thevenin impedance. (See Floyd, Section 20-5.)

   (a) What value is the resistive part of the load?

   (b) What value is the inductive part of the load?

3. Explain why the Thevenin circuit you tested in this experiment is equivalent only at 17 kHz.

4. Draw the Norton circuit for the circuit of Figure 32–2.

5. Compare the difference between applying Thevenin's theorem to a dc circuit with an ac circuit.

6. Assume the internal impedance of a particular source is 600–j100 Ω at a specified frequency. What must be the load impedance for maximum power transfer?

TECH TIP REVIEW:
Each of the following troubleshooting problems refers to the TECH TIP circuit shown in Figure 20–47, Schematic A, of the Floyd text. Predict the effect on the filter response for each of the following troubles.

1. Capacitor C_1 is open.

2. Capacitor C_2 is smaller than specified.

3. Resistor R_1 is larger than specified.

33 Integrating and Differentiating Circuits

OBJECTIVES:

After performing this experiment, you will be able to:

1. Explain how an *RC* or *RL* series circuit can act as an integrating or differentiating circuit.
2. Compare the waveforms for *RC* and *RL* circuits driven by a square-wave generator.
3. Determine the effect of a frequency change on the waveforms of pulsed *RC* and *RL* circuits.

READING:

Floyd, *Principles of Electric Circuits,* Sections 21–1 through 21–7 and 21–10

MATERIALS NEEDED:

One 100 mH inductor
One 0.01 μF capacitor
One 1000 pF capacitor
One 10 kΩ resistor
For Further Investigation: Second 10 kΩ resistor
Application Problem: One 1 kΩ potentiometer, one 7414 hex inverter with Schmitt trigger inputs

SUMMARY OF THEORY:

In mathematics, the word *integrate* means to sum. If we kept a running sum of the area under a horizontal straight line, the area would increase linearly. An example is the speed of a car. Let's say a car is traveling a constant 40 mi/h. In ½ h the car has traveled 20 mi. In 1 h the car has traveled 40 mi, and so forth. The car's rate is illustrated in Figure 33–1(a). Each of the three shaded areas under the rate curve represents 20 mi. If we plot the distance covered as a function of time, we obtain the graph in Figure 33–1(b). The graph of distance rises linearly with time, and represents the area (or integral) under the rate curve.

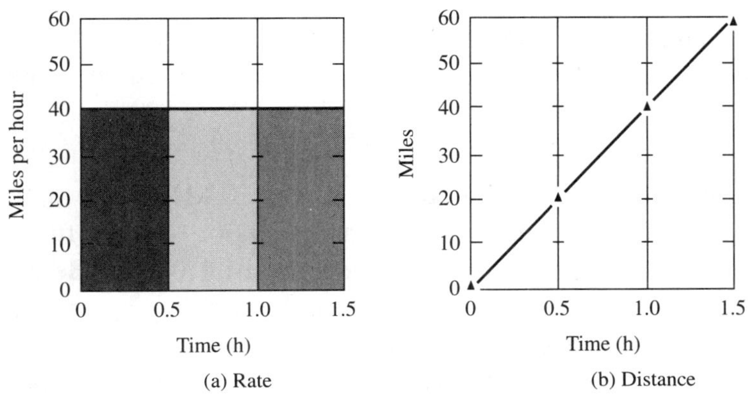

Figure 33–1

A similar situation exists when a capacitor starts to charge through a resistance toward a constant voltage. If the applied voltage remains a constant level, the voltage across the capacitor rises exponentially. However, if we examine a very short segment of this exponential rise, it appears to rise in a linear fashion similar to the mathematical integral. As long as the voltage change across the capacitor is small compared to the final voltage, the output will represent integration. An *integrator* is any circuit in which the output is proportional to the integral of the input signal. For certain applications, an *RC* circuit with the output across the capacitor is an integrator. It is easy to meet the condition that the voltage change is small compared to the final voltage. *If the RC time constant of the circuit is long compared to the period of the input waveform, then the waveform across the capacitor is integrated.*

The opposite of integration is *differentiation*. Differentiation is the process of finding the rate of change. *If the RC time constant of the circuit is short compared to the period of the input waveform, then the waveform across the resistor is differentiated.* A pulse waveform that is differentiated produces spikes at the leading and trailing edge, as shown in Figure 33–2. Differentiator circuits can be used to detect the leading or trailing edge of a pulse. Diodes can be used to remove either the positive or negative spike.

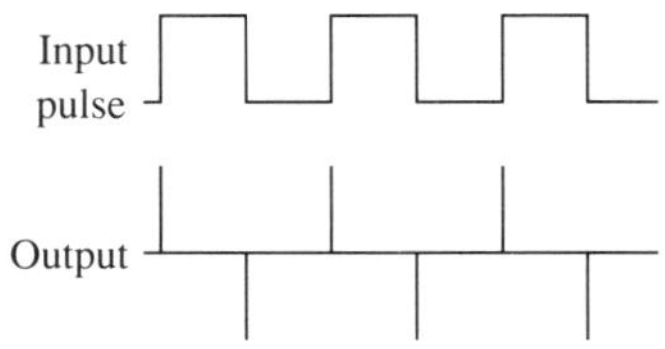

Figure 33–2

As you might suspect, an *RL* circuit can also be used as an integrator or differentiator. As in the *RC* circuit, the time constant for the *RL* integrating circuit must be long compared to the period of the input waveform and the time constant for the differentiator circuit must be short compared to the input waveform. The *RL* circuit will have waveforms similar to the *RC* circuit, except that the output signal is taken across the inductor for the differentiating circuit and across the resistor for the integrating circuit. The waveforms for both *RC* and *RL* integrating and differentiating circuits will be investigated in this experiment.

PROCEDURE:

1. Measure the value of a 100 mH inductor, a 0.01 μF and 1000 pF capacitor, and a 10 kΩ resistor. Record their values in Table 33–1. If it is not possible to measure the inductor or capacitors, use the listed values.

2. Construct the circuit shown in Figure 33–3. The 10 kΩ resistor is chosen to be large compared to the Thevenin impedance of the generator. Set the function generator for a 1.0 V_{pp} square wave with no load at a frequency of 1.0 kHz. Be sure the signal is dc coupled to the oscilloscope. You should observe that the capacitor fully charges and discharges at this frequency because the *RC* time constant is short compared to the period. On Plot 33–1, sketch the waveforms for the generator, the capacitor, and the resistor. Label voltage and time on your sketch. To look at the voltage across the resistor, use the difference technique described in Experiment 18.

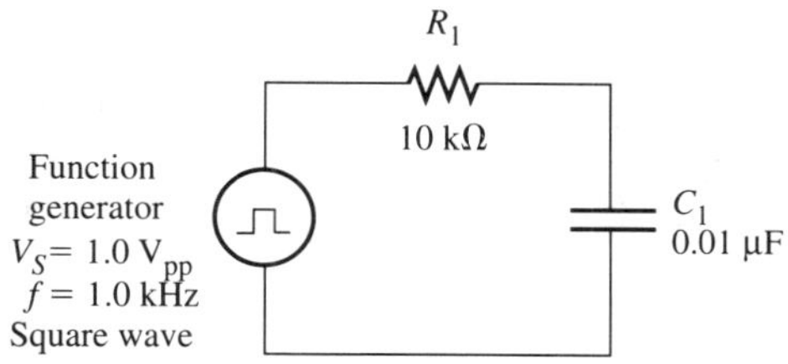

Figure 33–3

3. Compute the *RC* time constant for the circuit. Include the generator's Thevenin impedance as part of the resistance in the computation. Enter the computed time constant in Table 33–2.

4. Measure the *RC* time constant using the following procedure:
   (a) With the generator disconnected from the circuit, set the output square wave on the oscilloscope to cover five vertical divisions (0 to 100%).
   (b) Connect the generator to the circuit. Adjust the SEC/DIV and trigger controls to stretch capacitor charging waveform across the scope face to obtain best resolution.
   (c) Count the number of horizontal divisions from the start of the rise to the point where the waveform crosses 3.15 *vertical* divisions (63% of the final level). Multiply the number of *horizontal* divisions that you counted by the setting of the SEC/DIV control.

   Enter the measured *RC* time constant in Table 33–2.

5. Observe the capacitor waveform while you increase the generator frequency to 10 kHz. On Plot 33–2, sketch the waveforms for the generator, the capacitor, and the resistor at 10 kHz. Be sure to include the voltage and time on your sketch.

6. Temporarily change the function generator from a square wave to a triangle waveform. Describe in your report the waveform across the capacitor.

7. Change back to a square wave at the 10 kHz frequency. Then replace the 0.01 μF capacitor with a 1000 pF capacitor. Using the difference method, observe the waveform across the resistor. Describe the signal in your report.

8. Replace the 1000 pF capacitor with a 100 mH inductor. Keep the generator on the 10 kHz square wave and look at the signals on the generator, the inductor, and the resistor. On Plot 33–3, sketch the waveforms for each. Label voltage and time on your sketch.

FOR FURTHER INVESTIGATION:

The rate a capacitor charges is determined by the *RC* time constant. *R* and *C* are the equivalent series resistance and capacitance of the circuit. The *RC* time constant for the circuit in Figure 33–4 can be determined by applying Thevenin's theorem to the circuit to the left of points **A-A** on the circuit. The Thevenin resistance of the generator is part of the charging path and *should* be included. Note that the capacitor in this circuit is not charging to the generator voltage but to a voltage determined by the voltage divider consisting of R_1 and R_2. See if you can predict the time constant and the waveforms across each resistor. Investigate carefully the waveforms across each of the components in the circuit. Show that your observations satisfy Kirchhoff's voltage law at any instant in time.

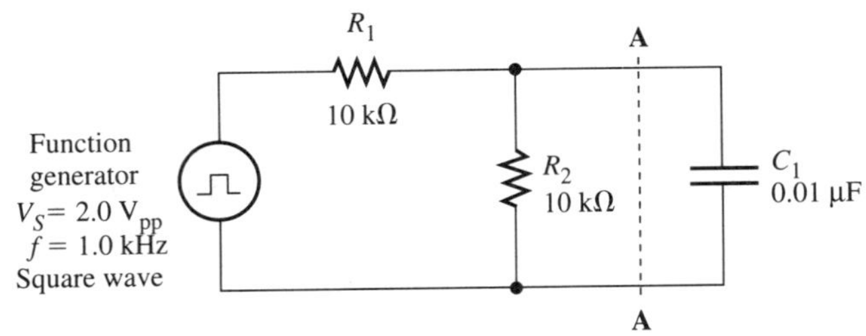

Figure 33–4

APPLICATION PROBLEM:

In step 5, you observed how a square wave can be converted into a triangle wave by using an integrator. A triangle wave can be generated in a simple integrating circuit by using a Schmitt trigger. A Schmitt trigger is a switching circuit with two thresholds for change. The switching level is dependent on whether the input signal is rising or falling. Consider the circuit shown in Figure 33–5(a). The charging and discharging of the capacitor is determined by the switching points of the Schmitt trigger. The input voltage is initially low and the output voltage is high (near 5.0 V). The capacitor begins to charge toward the higher output voltage. As the capacitor charges, the input voltage passes a trip point causing the input voltage to go high and the output voltage to go low. The capacitor begins to discharge toward the lower voltage until it passes the lower trip point causing the process to repeat.

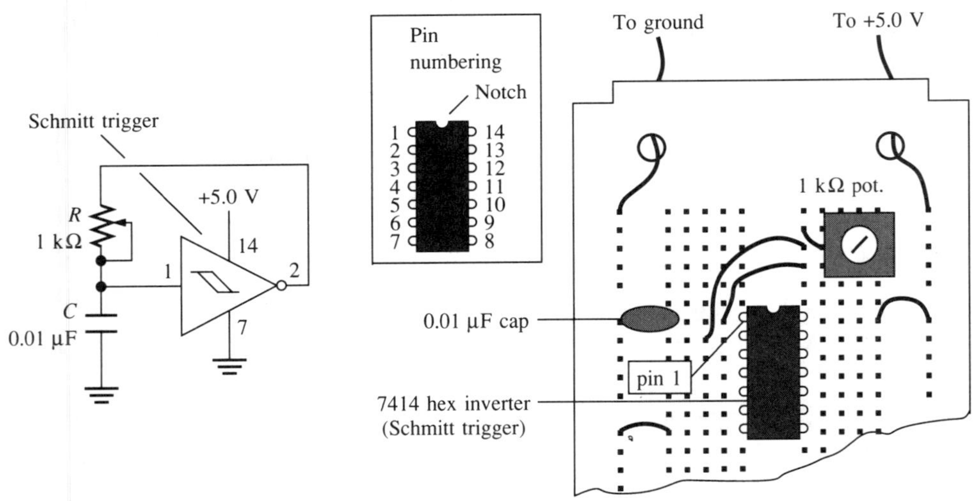

(a) Schematic

(b) Protoboard wiring

Figure 33–5

Construct the circuit and measure the waveform across the capacitor. The wiring is shown in Figure 33–5(b). Try varying R as you observe the capacitor voltage. What is the minimum and maximum frequency? Can you determine the threshold voltages of the Schmitt trigger by observing the output?

Report for Experiment 33

Name ____________
Date ____________
Class ____________

OBJECTIVES:

PROCEDURE:

DATA:

Table 33–1

	Listed Value	Measured Value
L_1	100 mH	
C_1	0.01 μF	
C_2	1000 pF	
R_1	10 kΩ	

Table 33–2

	Computed	Measured
RC time constant		

Step 2 and Step 5 (*RC* Circuits)

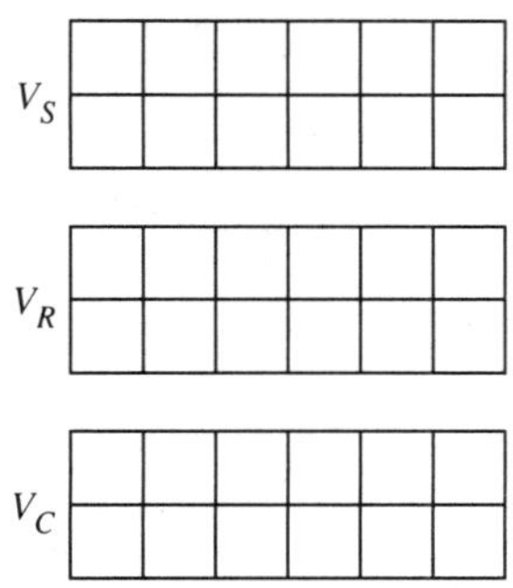

Plot 33–1

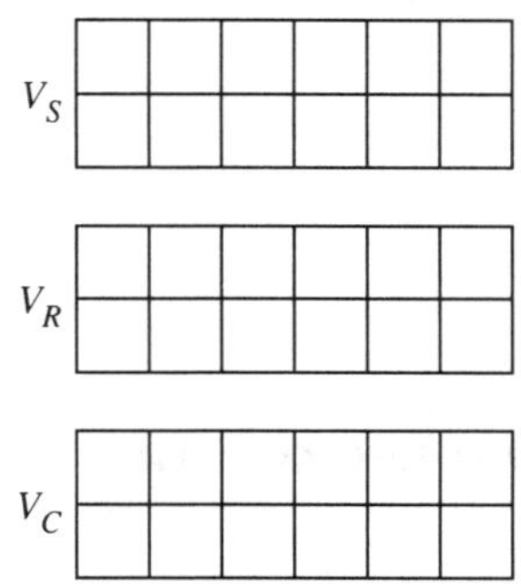

Plot 33–2

Step 6 (Triangle Waveform Results):

Step 7 (Square Wave with 1000 pF Capacitor):

Step 8 (Square Wave with 100 mH Inductor):

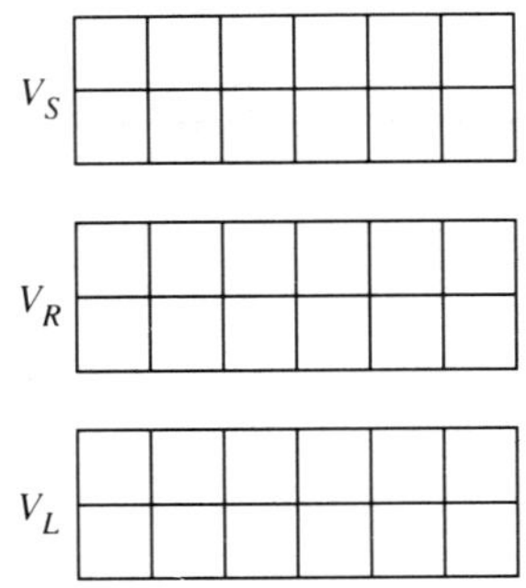

Plot 33–3

RESULTS AND CONCLUSION:

FURTHER INVESTIGATION RESULTS:

APPLICATION PROBLEM RESULTS:

EVALUATION AND REVIEW QUESTIONS:

1. (a) Explain why the Thevenin impedance of the generator was included in the calculated *RC* time constant measurement in step 3.

   (b) Suggest how you might find the value of an unknown capacitor using the *RC* time constant.

2. (a) Compute the % difference between the measured and computed *RC* time constant.

   (b) List some factors that affect the accuracy of the measured result.

3. What accounts for the change in the capacitor voltage waveform as the frequency was raised in step 5?

4. (a) Draw an *RC* integrating circuit and an *RC* differentiating circuit.

   (b) Draw an *RL* integrating circuit and an *RL* differentiating circuit.

5. Assume you had connected a square wave to an oscilloscope but saw a signal that appeared to be differentiated as illustrated in Figure 33–2. What could account for this effect?

6. What type of circuit is required in order to change a square-wave into a triangle-wave output?

TECH TIP REVIEW:
Each of the following questions refers to the TECH TIP circuit shown in Figure 21–56 of the Floyd text. Your supervisor wants to consider the cost trade-off between the circuit shown and the possibility that the delay times (listed in Table 21–1 of Floyd's text) could be implemented using a variable threshold circuit instead of a fixed 3.5 V threshold. To ensure reliability, the threshold voltage should be no less than 0.8 V and no greater than 4.2 V.

1. Show that it is possible to obtain the required delay times with the range of threshold voltages given in the preceding problem statement.

2. For the threshold conditions given in problem statement, draw a circuit using the 47 kΩ resistor specified in Floyd's text and one of the standard value capacitors listed in Activity 1. Indicate on your drawing the value of the capacitor needed.

Appendix A: Chapter Summaries

The following summaries include key ideas from Floyd's text organized by chapter and section for ready reference to the text.

CHAPTER 1 COMPONENTS, QUANTITIES, AND UNITS

Section 1–1 Components and Measuring Instruments

1. Basic electronic components are the resistor, capacitor, inductor, and transformer. The characteristics of circuits containing these components is the subject of this text.
2. Important measuring instruments for basic electrical quantities include the voltmeter, ammeter, ohmmeter, oscilloscope, and wattmeter.

Section 1–2 Electrical and Magnetic Units

Electrical and magnetic quantities are measured with *SI* units *(Systeme International* in French). These quantities will be defined in the text as they are introduced.

Section 1–3 Scientific Notation

Scientific notation uses powers of ten to express numbers. It is convenient for representing very large or very small numbers. In scientific notation, a number is written as a number between 1 and 10 times a power of ten.

Section 1–4 Metric Prefixes

1. Metric prefixes are another way of expressing large and small numbers. In electrical work, the most useful metric prefixes are those that are multiples of three. These include pico (10^{-12}), nano (10^{-9}), micro (10^{-6}), milli (10^{-3}), kilo (10^{3}), mega (10^{6}), giga (10^{9}), and tera (10^{12}).
2. Engineering notation uses power of 10 notation with the requirement that the exponent be a multiple of 3. A number written in engineering notation can be directly converted to a number written with a metric prefix and vice-versa.

Section 1–5 Metric Unit Conversions

Numbers written with metric prefixes can be converted to a different prefix by moving the decimal. If the number is larger, the metric prefix will be smaller—if the number is smaller, the metric prefix will be larger.

Section 1–6 Introduction to PSpice

PSpice is a text-based simulation program for electronic circuits. Components and sources are identified by the first letter of their name. Component values are assigned with fixed or scientific numbers or with metric prefixes.

CHAPTER 2 VOLTAGE, CURRENT, AND RESISTANCE

Section 2–1 Atoms

1. The basic building block of elements is the atom. According to the Bohr model, atoms consist of a positive nucleus containing protons and neutrons, with negatively charged electrons in orbits around the nucleus. Elements differ from one another by the number of protons in the nucleus, which is their atomic number. The elements are listed on the periodic table in order of their atomic number.

2. Electrons are arranged in shells around the nucleus. Shells that are further from the nucleus can hold more electrons.
3. Metals form good conductors because their outer shell electrons are easily freed. Nonmetals form insulators because they have very few free electrons. Certain materials, especially silicon and germanium, form semiconductors, which are between conductors and insulators.

Section 2–2 Electrical charge
Electrical charge, Q, is defined by the coulomb (C), which is a very large number of electrons (6.25×10^{18}).

Section 2-3 Voltage
1. Voltage, V, is defined by *two* points and is measured in volts (V). One volt is defined as the potential difference between two points when one joule of energy is required to move one coulomb of charge from one point to the other.
2. Sources of voltage include batteries, power supplies, solar cells, and generators.

Section 2-4 Current
Current, I, is the rate that charge flows and is measured in amperes (A). One ampere is defined as the current that is flowing when 1 coulomb (1 C) of charge passes a given point in 1 second (1 s).

Section 2-5 Resistance
1. Resistance, R, is the opposition to current flow and is measured in ohms (Ω). One ohm (1 Ω) of resistance exists when one ampere (1 A) of current flows in a material when one volt (1 V) is applied across the material.
2. Conductance, G, is the reciprocal of resistance and is measured in siemens (S).
3. Resistors are components designed to have a specific amount of resistance. The value and tolerance is determined from the resistor color code.
4. Variable resistors are designed to change resistance. Some resistors change automatically with temperature (thermistors) or light (photoconductive cells).

Section 2-6 The Electric Circuit
1. A basic circuit consists of a voltage source, a path for current, and a load.
2. Switches are used to open or close the circuit path. Switches are described by the number of poles (moveable contacts) and the number of throws (contacts) for each pole. Pushbutton and rotary switches are other types.
3. Circuit breakers and fuses are designed to open if current exceeds a predetermined amount.
4. Wires are arranged by standard gage numbers called American Wire Gage (AWG). Larger diameter wire has a smaller gage number. The resistance of wire is proportional to its resistivity and length and inversely proportional to its area.
5. Voltage must be measured between two points. Ground is the reference point in electric circuits and has a potential of 0 V with respect to other points in the circuit.

Section 2-7 Basic Circuit Measurements
1. The multimeter is a combination instrument that can measure voltage, resistance, and current. Multimeters are available as analog meters (the VOM) or digital meters (the DMM).
2. Ammeters are connected in series with the current to be measured. (Note that an ammeter is *never* connected directly across a voltage source.) Voltmeters and ohmmeters are connected in parallel. An ohmmeter should never be used in an energized circuit.
3. The resolution of a meter is the smallest increment of a quantity the meter can measure. Accuracy is the degree to which a measured value represents the true or accepted value of a quantity. Meter accuracy is determined by the internal circuitry.

Section 2-8 TECHnology Theory Into Practice
This section illustrates practical measurements with meters in a basic dc circuit.

Section 2-9 PSpice Circuit Connections
Pspice component information is given by the format:

Component name node 1 node 2 . . . node n Component value

CHAPTER 3 OHM'S LAW

Section 3–1 Ohm's Law
Ohm's law expresses the relationship between voltage, current and resistance in a circuit. Mathematically it is written as $I = \frac{V}{R}$ where I stands for the current in amperes, V is the voltage in volts, and R is the resistance in ohms.

Section 3–2 Calculating Current

1. For calculating current in a dc circuit, Ohm's law is written as $I = \frac{V}{R}$. Metric prefixes are useful for computing values using Ohm's law.
2. When resistance is expressed in ohms (Ω) and voltage in volts (V), the current will be in amperes. Resistance is frequently expressed as thousands (kilo) or millions (mega). When resistance is in kilohms (kΩ), and voltage is in volts (V), current will be in milliamperes (mA). When resistance is in megohms (MΩ), and voltage is in volts (V), current will be in microamperes (μA).

Section 3–3 Calculating Voltage
For calculating voltage in a dc circuit, Ohm's law is written as $V = IR$.

Section 3–4 Calculating Resistance
For calculating resistance in a dc circuit, Ohm's law is written as $R = \frac{V}{I}$.

Section 3–5 The Relationship of Current, Voltage, and Resistance
1. For a constant value of resistance, the relationship between current and voltage is linear and the graph of current versus voltage produces a straight line.
2. If voltage is held constant and resistance is varied, the plot of current versus resistance shows an inverse relationship.

Section 3–6 TECHnology Theory Into Practice
This section illustrates a practical application of Ohm's law with a resistor box.

Section 3–7 PSpice Files
PSpice files are created in the circuit editor and are given a .CIR extension. The first statement in the file is the title. Typically, the last statement is the .END directive, which signals the end of file. When PSpice is run, an output file is created with a .OUT extension of the original file name.

CHAPTER 4 POWER AND ENERGY

Section 4–1 Power and Energy
Work is done when a force acts over a distance. In the SI system, the unit for measuring work or

energy is the joule. The rate at which energy is converted to another form is power, which is measured in joules per second or watts.

Section 4–2 Power in an Electric Circuit
Power in dc circuits can be determined from the product of current and voltage. Substituting Ohm's law, power can be obtained from any of three equivalent equations:

$$P = I^2R = VI = \frac{V^2}{R}$$

Section 4–3 Resistor Power Ratings
The power rating of a resistor is the maximum power it can dissipate. The rating is related to the physical size and shape of the resistor. If the power rating is exceeded, typically the resistor is "burned out," making it an open.

Section 4–4 Energy Loss and Voltage Drop in Resistance
Energy in resistances is converted from electrical form to heat.

Section 4–5 Power Supplies
1. Power supplies produce voltage and supply the voltage and current to a load. The product of current and output voltage is the power supplied to the load.
2. The ampere-hour rating of a battery is proportional to its energy capacity. The ampere-hours indicate the maximum product of the current times the time at the rated output voltage of the battery.
3. Power supply efficiency is the ratio of the output power to the input power.

Section 4–6 TECHnology Theory Into Practice
This section illustrates a practical application for solving power dissipation in a resistor.

Section 4–7 PSpice DC Analysis
The .DC control statement allows you to sweep a source voltage or current parameter over a range of values.

CHAPTER 5 SERIES CIRCUITS

Section 5–1 Resistors in Series
A *series circuit* has only one closed path for current. When resistors (or other components) are connected together, they are said to be *in series* if there is only one path.

Section 5–2 Current in a Series Circuit
Current is the same throughout a series circuit.

Section 5–3 Total Series Resistance
1. When resistors are connected in series, the total resistance is equal to the sum of the individual resistors. That is:

$$R_T = R_1 + R_2 + R_3 + \ldots R_n$$

where n is a positive integer equal to the number of resistors in the string.

2. When a series string has equal value resistors, the total resistance is

$$R_T = nR$$

where R is the value of one of the equal value resistors.

Section 5–4 Ohm's Law in Series Circuits
1. In a series circuit with one source, the current can be found by dividing the source voltage by the total resistance. This current is the same in all components.
2. The voltage drop across any resistor is the current times the resistance.

Section 5–5 Voltage Sources in Series
When voltage sources are connected in series, the total voltage is the algebraic sum of the individual source voltages. The polarities must be taken into account when finding the algebraic sum.

Section 5–6 Kirchhoff's Voltage Law
Kirchhoff's voltage law states that the algebraic sum of voltages around any closed path is equal to zero.

Section 5–7 Voltage Dividers
The voltage divider rule is developed for series circuits. It can be stated as: **The voltage drop across a resistor in a series circuit is equal to the ratio of the resistor to the total resistance, multiplied by the source voltage.** This rule can be applied to a combination of resistances in a series circuit. The mathematical statement of the voltage divider rule is

$$V_x = \frac{R_x}{R_T}V_s$$

Section 5–8 Power in a Series Circuit
The total power dissipated in a series resistive circuit is the sum of the power dissipated in each resistor. The total power can be found from the total current multiplied by the total voltage or other relationships found by substituting Ohm's law and can be expressed as:

$$P_T = I^2R_T = V_sI = \frac{V_S^2}{R_T}$$

Section 5–9 Circuit Ground
1. Ground is the point in a circuit that is the reference point for voltage measurements. When a voltage is given with a single lettered subscript, such as V_A, the reference point is ground.
2. Voltage can be measured across an ungrounded component. When a voltage is given as V_{AB}, it is measured between points A and B of the circuit.

Section 5–10 Troubleshooting
1. The most common problem in a series circuit is an open circuit. When this occurs, the source voltage is found across the open and no current flows in the circuit.
2. A short circuit means that two normally separated parts of a circuit are connected together by an unwanted conductor.

Section 5–11 TECHnology Theory Into Practice
This section illustrates a voltage divider on a printed circuit board.

Section 5–12 PSpice DC Analysis
1. The .OP directive is a statement that causes PSpice to produce operating information (discussed later). In a series circuit, it will show the node voltages.
2. The .PLOT directive shows data as a textual graph or plot.

CHAPTER 6 PARALLEL CIRCUITS

Section 6–1 Resistors in Parallel
When two or more resistors are individually connected across the same points, they are in parallel with each other. A *parallel circuit* includes a voltage source and has more than one path for current. Each current path in a parallel circuit is called a *branch*.

Section 6–2 Voltage Drop in Parallel Circuits
In a parallel circuit, the voltage across all components is the same as the source voltage.

Section 6–3 Kirchhoff's Current Law
Kirchhoff's current law states that the sum of the currents entering a junction (node) is equal to the sum of the currents out of that junction.

Section 6–4 Total Parallel Resistance
1. In a parallel circuit, the *reciprocal* of the total resistance is the sum of the *reciprocals* of the individual resistors. In equation form this is

$$\frac{1}{R_T} = \frac{1}{R_1} + \frac{1}{R_2} + \frac{1}{R_3} + \ . \ . \ . \ + \frac{1}{R_n}$$

This is equivalent to saying the total conductance is equal to the sum of the individual conductances ($G_T = G_1 + G_2 + G_3 + \ . \ . \ . \ G_n$).
2. For the special case of two resistors in parallel, the total resistance can be found by the product-over-sum rule. That is

$$R_T = \frac{R_1 R_2}{R_1 + R_2}$$

3. For convenience, parallel resistors are often indicated by two parallel lines between them; for example $R_1//R_2$ means that R_1 is in parallel with R_2.

Section 6–5 Ohm's Law in Parallel Circuits
1. In parallel circuits, the voltage across each resistor is the same as the source voltage. Therefore, to find the current in any resistor, divide the source voltage by the resistance.
2. The total current in a parallel circuit is the source voltage divided by the total resistance.

Section 6–6 Current Sources in Parallel
The total current produced by current sources in parallel is equal to the algebraic sum of the individual current sources.

Section 6–7 Current Dividers

1. The current divider rule is derived from Ohm's law and applied to parallel circuits. In general, the current in a resistor is equal to the total current multiplied by the ratio of the total parallel resistance divided by the resistor. That is:

$$I_x = \frac{R_T}{R_x} I_T$$

2. For the special case of two resistors, the current divider rule can be written

$$I_1 = \left(\frac{R_2}{R_1 + R_2}\right) I_T \text{ and } I_2 = \left(\frac{R_1}{R_1 + R_2}\right) I_T$$

Section 6–8 Power in Parallel Circuits

The total power dissipated in a parallel resistive circuit is the sum of the power dissipated in each resistor. The total power can be found from the total current multiplied by the total voltage or other relationships found by substituting Ohm's law and expressed the same as the series case. That is

$$P_T = I_T^2 R_T = V_S I_T = \frac{V_S^2}{R_T}$$

Section 6–9 Examples of Parallel Circuit Applications

Parallel circuits are used in residential wiring, automobiles, and other applications where a voltage that is independent of other loads is needed.

Section 6–10 Troubleshooting

An open branch in a parallel circuit will have no effect on the source voltage or current in other branches but will reduce the total current in the circuit. The open branch may be found by a current measurement. A resistance measurement may also reveal the open, provided care is taken to isolate the path.

Section 6–11 TECHnology Theory Into Practice

This section illustrates a practical parallel circuit in a milliammeter range switch.

Section 6–12 PSpice DC Analysis

A parallel circuit file has the voltage source and resistors connected to the same set of nodes. The source voltage can be varied by using a .DC statement; the current in a given resistor can be plotted as a function of the varying voltage with the .PLOT statement.

CHAPTER 7 SERIES-PARALLEL CIRCUITS

Section 7–1 Identifying Series-Parallel Relationships

Analysis of combination circuits usually begins by identifying series-parallel relationships. When components are in series, there is only one path for current and the identical current will flow in each. When components are in parallel, both ends are joined together and the voltage is identical across them.

Section 7–2 Analysis of Series-Parallel Circuits
After identifying series-parallel relationships, circuit analysis is frequently done by forming equivalent circuits. Equivalent circuits are formed by replacing two or more resistors by a single equivalent resistance. Equivalent circuits may be formed more than once to reduce the circuit to a single total resistance. The total current can then be found from Ohm's law. The analysis continues by applying basic circuit laws until all currents and voltages of interest have been found.

Section 7–3 Voltage Dividers with Resistive Loads
A loaded voltage divider is a basic resistive combination circuit. Currents and voltages can be solved by applying equivalent circuit analysis methods described in Section 7–2.

Section 7–4 Loading Effects of a Voltmeter
A voltmeter has an internal resistance that depends on the meter's sensitivity factor and the range selected. This resistance is placed in parallel with the circuit to be measured and causes the circuit's voltage to drop. This effect, called *meter loading,* is more serious in high resistance circuits.

Section 7–5 Ladder Networks
A resistive ladder network is a combination circuit that can be analyzed by reducing the network into equivalent circuits. One application of ladders is the *R*/2*R* network used in digital to analog (D/A) conversion.

Section 7–6 The Wheatstone Bridge
The Wheatstone bridge is a sensitive circuit used in measurement applications. The bridge is balanced when the output voltage (or current) is zero. This occurs when the resistor ratios on each side of the bridge are matched.

Section 7–7 Troubleshooting
Troubleshooting requires the ability to apply knowledge of circuit laws and logic to determine why a circuit is not operating correctly. Typical problems with circuits are shorts, opens, or wrong values of components.

Section 7–8 TECHnology Theory Into Practice
This section illustrates a loaded voltage divider.

Section 7–9 PSpice DC Analysis
No new statements are introduced. An example of a series-parallel resistive combination circuit is given.

CHAPTER 8 CIRCUIT THEOREMS AND CONVERSIONS

Section 8–1 The Voltage Source
An ideal voltage source provides a constant voltage regardless of the load. This implies that the internal resistance is zero. Practical voltage sources are drawn with an ideal voltage source in *series* with the internal resistance of the source, which is (normally) small. It is this internal resistance that causes loading of the source.

Section 8–2 The Current Source
An ideal current source provides a constant current regardless of the load. This implies that the internal resistance is infinitely large. Practical current sources are drawn with an ideal current source in *parallel* with the internal resistance of the source, which is (normally) very large. It is this internal resistance that causes loading of the source.

Section 8–3 Source Conversions

1. A practical voltage source is converted to a current source by applying Ohm's law to a shorted output on the voltage source. The value of the current source is then

$$I_S = \frac{V_S}{R_S}$$

2. A practical current source is converted to a voltage source by applying Ohm's law to an open output on the current source. The value of the voltage source is

$$V_S = I_S R_S$$

3. The series resistance of the voltage source is the same as the parallel resistance of the current source for both conversions.

Section 8–4 The Superposition Theorem

When a linear circuit has more than one source, the currents and voltages can be found by computing each source independently, and algebraically combining the results. When computing a particular source, all other sources are replaced with their internal resistance.

Section 8–5 Thevenin's Theorem

Any two-terminal, linear network can be replaced with an ideal voltage source (called the *Thevenin voltage*) in series with a resistor (called the *Thevenin resistance*). The Thevenin voltage is equal to the open circuit voltage of the original circuit. The Thevenin resistance is the resistance seen at the two terminals of the original circuit, with all sources replaced with their internal resistances.

Section 8–6 Norton's Theorem

Any two-terminal, linear network can be replaced with an ideal current source (called the *Norton current*) in parallel with a resistor (called the *Norton resistance).* The Norton current is equal to the current that would flow in a short placed on the output terminals of the original circuit. The Norton resistance is the resistance seen at the two terminals of the original circuit, with all sources replaced with their internal resistances (same as the Thevenin resistance).

Section 8–7 Millman's Theorem

Millman's theorem allows several parallel Thevenin sources to be reduced to a single source. It is applied by converting the parallel voltage sources to current sources and algebraically adding the current sources. The process of converting the voltage sources to current sources creates parallel resistors which are combined into one equivalent resistance. The final equivalent source is usually drawn as a new Thevenin circuit.

Section 8–8 Maximum Power Transfer Theorem

When a circuit is connected to a load, maximum power is transferred to the load when the load resistance is equal to the source resistance of the circuit. (This theorem assumes that the source resistance cannot be varied).

Section 8–9 Delta-to-Wye (Δ-to-Y) and Wye-to-Delta (Y-to-Δ) Conversions

1. A delta (Δ) network is a three terminal network drawn in the shape of a triangle. A Y network is drawn in the shape of the letter Y. These networks are commonly used in three phase systems, discussed later.

2. A resistive-Δ or resistive-Y network can be converted from one to the other with the equations given in the text. This allows certain complex circuits, such as the loaded Wheatstone bridge, to be solved.

Section 8–10 TECHnology Theory Into Practice
This section illustrates the analysis of a practical Wheatstone bridge circuit.

Section 8–11 PSpice DC Analysis
1. A Thevenin circuit may be computed with PSpice by replacing the load resistor with an *extremely* large value to simulate an open. The voltage is computed across this load and represents the Thevenin voltage.
2. A Norton circuit may be found by replacing the load resistor with a short and computing the current in the short. This represents the Norton current.
3. The Thevenin and Norton resistances are equal. To find this resistance, divide the Thevenin voltage by the Norton current.
4. The Thevenin and Norton problems can be run in a single pass with a multiple circuit file.

CHAPTER 9 BRANCH, MESH, AND NODE ANALYSIS

Section 9–1 Branch Current Method
1. A *loop* is a complete current path within a circuit. A *node* is a point in a circuit where two or more elements join. A *branch* is a path that connects two nodes. The maximum number of unknown currents in a circuit is equal to the number of branches.
2. To use the branch current method, assign a polarity across each resistor consistent with an assumed direction of current. Write one equation for each unknown current by writing Kirchhoff's voltage law around each closed loop (watching signs). Next write Kirchhoff's current law at a minimum number of nodes to give the required number of equations (one equation for each unknown current). The simultaneous equations are then solved by algebra to find the currents.

Section 9–2 Determinants
Determinants are a system of solving simultaneous equations by application of properties of matrixes.

Section 9–3 Mesh Current Method
The mesh current method for solving a circuit uses loop equations and application of the superposition theorem. Kirchhoff's voltage law is written around each closed loop, resulting in one equation for each loop. The resulting simultaneous equations are solved for the loop currents. In components with more than one loop current, the total current is the algebraic sum of the loop currents.

Section 9–4 Node Voltage Method
The node voltage method for solving the voltage at one or more nodes is based on Kirchhoff's current law. First, the current law is written for a principal node (three or more components). A reference point is selected and the node currents are then converted (using Ohm's law) to a pair of node voltages divided by a resistance. The resulting equation (or equations) is solved for the unknown node voltage (or voltages).

Section 9–5 TECHnology Theory Into Practice
This section illustrates application of analysis methods to a voltage with both positive and negative source voltages.

Section 9–6 PSpice DC Analysis
With multiple sources in a circuit, an analysis control statement such as .DC is necessary before the .PRINT statement in order to find the current in a specific resistor (see the PSpice example in Experiment 11 also).

CHAPTER 10 MAGNETISM AND ELECTROMAGNETISM

Section 10–1 The Magnetic Field
1. Magnetic fields are represented by invisible lines of force drawn from the north pole to the south pole of a magnet. When magnetic materials are brought near a magnet, the field lines are altered and tend to pass through the magnetic material. A group of lines is called *magnetic flux,* ϕ, measured in webers (Wb). A weber represents a large unit of magnetic flux (1×10^8 lines).
2. *Magnetic flux density* is the flux per unit area perpendicular to the flux lines and is defined by the formula

$$B = \frac{\phi}{A}$$

where B = flux density in tesla, ϕ = flux in webers, and A = area in square meters.
3. The magnetic properties of ferromagnetic materials is explained by domain theory.

Section 10–2 Electromagnetism
1. Conventional current flow (+ to −) in a conductor creates a magnetic field surrounding the conductor given by the right hand rule.
2. *Permeability,* μ, is a measure of the ease of establishing a magnetic field in a material. *Relative permeability,* μ_r, is the ratio of the permeability of a material to that of a vacuum.
3. *Reluctance,* $\mathcal{R}$, is the opposition to the establishment of a magnetic field in a material.
4. *Magnetomotive force* is the product of current, I, and number of turns of wire, N. Magnetomotive force is not a true "force" in the physics sense but establishes the magnetic field. The defining formula for magnetomotive force is

$$F_m = NI$$

where F_m = magnetomotive force in ampere-turns, N = number of turns, I = current in amperes
5. Magnetic circuits have an analogy with Ohm's law for electrical circuits: flux is analogous to current, magnetomotive force is analogous to voltage and reluctance is analogous to resistance. Ohm's law for magnetic circuits is:

$$\phi = \frac{F_m}{\mathcal{R}}$$

Section 10–3 Electromagnetic Devices
1. A *solenoid* uses an electromagnet to move a plunger when an electric current flows in its coil. The plunger is returned to its rest position by a spring when no current flows.

2. A *relay* uses an electromagnet to control switch contacts. Switch contacts are marked for the unenergized position as either normally closed (NC) or normally open (NO).
3. A *speaker* uses an electromagnet that interacts with a permanent magnet to rapidly drive a flexible cone in and out depending on the current in the electromagnet.
4. Analog meters use the interaction of an electromagnet and a permanent magnet to move a needle an amount proportional to the current in the electromagnet. The force due to the electromagnet is opposed by a spring; when the forces are balanced, the needle stops.

Section 10–4 Magnetic Hysteresis

1. *Magnetizing force, H,* (also referred to as magnetic field strength) is magnetomotive force per length of a material. The magnetizing force is defined by the equation

$$H = \frac{F_m}{l}$$

where H = magnetizing force in ampere-turns per meter, F_m = magnetomotive force in ampere-turns, and l = length of the material.
2. When current is increased in a coil that is wound around a magnetic material, the magnetizing force, H, increases proportionately. At first, the flux density, B, increases also. When the flux reaches some point, further increases in H do not affect the flux density. At this point the material is said to be in *saturation*. Decreasing the current to zero does not necessarily cause the flux to return to zero. This phenomenon is called *retentivity,* a property that depends on the material. The plot of B versus H for a material is called the *hysteresis curve*.

Section 10–5 Electromagnetic Induction

1. When a conductor is moved in a direction perpendicular to a magnetic field (or the field is moved relative to the conductor), a voltage is induced in the conductor. When the circuit is completed, an induced current flows in the conductor.
2. An induced current creates a magnetic field that interacts with the original magnetic field, creating a force on the conductor. This force is perpendicular to the motion of the conductor and to the original magnetic field. Notice that the three quantities, namely, the magnetic field, the induced current, and the force, are all at right angles to each other as illustrated in Figure 10–29 of the text.
3. Faraday's law is a mathematical statement of the principle of electromagnetic induction. It is written

$$v_{\text{ind}} = N\left(\frac{d\phi}{dt}\right)$$

where v_{ind} is the induced voltage in volts, N = number of turns, $\frac{d\phi}{dt}$ = rate of flux change, in webers per second.

Section 10–6 Applications of Electromagnetic Induction

1. A *crankshaft position sensor* uses magnetic induction to produce a voltage when a tab moves through an air gap of a magnet.
2. A *dc generator* produces current in an external circuit when a coil of wire is rotated by an external force in a magnetic field.

Section 10–7 TECHnology Theory Into Practice
This section illustrates a practical magnetic device, the relay, in an alarm circuit.

Section 10–8 Introduction to Probe
Probe is a postprocessor program that works with a .DAT file created by PSpice. PSpice creates the .DAT file when a .PROBE statement is included in the PSpice circuit file. All currents and voltages are included in the .DAT file for use in plotting. The user selects the dependent variable (voltage or current) to plot from a menu. The independent variable on the plot is a quantity that has been varied in PSpice, and is typically the input voltage. In ac analysis (covered later), frequency can be the independent variable.

CHAPTER 11 INTRODUCTION TO ALTERNATING CURRENT AND VOLTAGE

Section 11–1 The Sine Wave
1. A *sine wave* in electrical work has the shape of the trigonometric function but can represent an alternating current or voltage. It is the most basic electrical wave, formed in rotating machinery and oscillator circuits.
2. The *period, T,* of a sine wave is the time required for one complete cycle (a positive and a negative alteration). It is measured in seconds.
3. The *frequency, f,* of a sine wave is the number of cycles in one second measured in Hertz.
4. The period is the reciprocal of the frequency, and the frequency is the reciprocal of the period. That is

$$f = \frac{1}{T} \qquad T = \frac{1}{f}$$

Section 11–2 Sine Wave Voltage Sources
1. Sine waves can be generated by ac generators. The frequency of the sine wave is the product of the revolutions per second times the number of pole pairs.
2. Sine waves can also be generated by electronic signal generators that can provide a wide range of frequencies and amplitudes.

Section 11–3 Voltage and Current Values of Sine Waves
1. The *instantaneous value* of a sine wave is the voltage or current value that is indicated at a specific point in time.
2. The *peak* value of a sine wave is the voltage or current value that is the maximum value measured from the center (zero) of the wave.
3. The *peak-to-peak* value of a sine wave is twice the peak value.
4. The *average* value of a sine wave is calculated over a half-cycle and is 0.637 times the peak value.

Section 11–4 Angular Measurement of a Sine Wave
1. There are 2π radians in a revolution and 360° in a revolution. Therefore, one radian is equal to 57.3°.
2. The points along the *x*-axis of a sine wave can be assigned angular values (that correspond to the position of a rotating vector). The *phase* (or phase angle) specifies the position of the sine wave relative to some reference. When a sine wave is shifted to the left or right of the reference, there is a *phase shift*.

3. Lead and lag refer to the phase shift of a sine wave relative to a reference. A sine wave *leads* the reference if it occurs earlier (up to 180°); it *lags* if it occurs later.

Section 11–5 The Sine Wave Formula
The sine wave formula is a mathematical expression for a sine wave. It is:

$$y = A \sin (\theta \pm \phi)$$

where y is the instantaneous y value, A is the amplitude (measured between the center and the peak), θ = angle along the x-axis for which y is computed, ϕ = phase shift between sine wave and the reference. Note that the sign of ϕ is positive if the wave leads the reference and is negative if the wave lags the reference.

Section 11–6 Ohm's Law and Kirchhoff's Law in AC Circuits
Ohm's law and Kirchhoff's laws can be applied to ac resistive circuits in the same manner as they were applied to dc circuits. The voltage and current need to be expressed in a consistent set of values (all rms, all peak, etc.).

Section 11–7 Superimposed DC and AC Voltages
The superposition theorem applies to circuits with superimposed dc and ac voltages. Each source is treated independently and the results are combined.

Section 11–8 Nonsinusoidal Waveforms
1. A *pulse* is an electrical signal that makes a rapid transition from a baseline to some level, remains at that level for an interval of time and returns rapidly to the baseline. Pulses can be positive or negative going. The *leading-edge* and *trailing-edge* are specified by *rise time* (time from 10% to 90% of the amplitude) and *fall time* (time from 90% to 10% of the amplitude). *Pulse width* is specified at the 50% level.
2. Repetitive pulses are periodic waveforms that are characterized by their *pulse repetition frequency* (PRF), *period* (T), and *duty cycle* (pulse width divided by period). Duty cycle is usually expressed as a percentage.
3. Other nonsinusoidal waveforms include *square waves* (a special case of a pulse with a 50% duty cycle), *triangular waveforms*, and *sawtooth waveforms*. Triangular and sawtooth waveforms are characterized by *ramps*.
4. Repetitive nonsinusoidal waveforms have a *fundamental* frequency equal to the basic repetition rate of the waveform and *harmonics* that are multiples of this fundamental frequency.

Section 11–9 Oscilloscope Measurements
The oscilloscope is a basic measuring tool for viewing a voltage-versus-time graph of waveforms. Briefly, there are four basic sections common to virtually all oscilloscopes. These are the *display, vertical, horizontal,* and *trigger* sections. The display section has controls to find the beam, change its intensity, and focus it. The vertical controls vary the sensitivity of the y-axis and allow the user to change the vertical position of the signal, and select the channel. The horizontal controls vary the rate at which the beam moves across the screen. The trigger controls select a source and determine when the beam starts to assure a stable display. Operation of an oscilloscope is a very important skill for electronics technology; you are urged to learn this skill by practice.

Section 11–10 TECHnology Theory Into Practice
This section illustrates a measurement of waveforms within a radio receiver.

Section 11–11 Introduction to AC Analysis with PSpice and Probe

1. AC sources are specified with the following format:

<source name (starting with V or I)> <first node> <second node>
+AC <Peak Value> <Phase shift> <Transient values>

2. For an AC sinusoidal source, the specification for the transient value is SIN(<DC offset> <Peak value> <Frequency>)
3. The .TRAN control statement specifies how circuit values change with time.

CHAPTER 12 PHASORS AND COMPLEX NUMBERS

Section 12–1 Introduction to Phasors

1. A vector is a quantity with magnitude and direction. A phasor is a graphic representation of the magnitude and angular position of a time-varying quantity (a "rotating vector"). The length of the phasor arrow represents the magnitude of a quantity (such as voltage or impedance). The angle is represented by the direction the arrow is drawn and is with respect to some reference quantity (the reference is frequently the voltage in a parallel circuit or the current in a series circuit).
2. A phasor diagram can be used to show the relationship of two or more sine waves of the same frequency.
3. One cycle of a sine wave is traced out when a phasor is rotated through 360 degrees. The period and frequency are related to the velocity of rotation of the phasor. The velocity of rotation is called the angular velocity and is designated ω (the Greek letter omega).
4. The time required for a phasor to go through 2π radians is the period of the sine wave. Since the phasor rotates through 2π radians in a time equal to the period T, the angular velocity can be expressed in terms of either the period or frequency as

$$\omega = \frac{2\pi}{T} = 2\pi f$$

Section 12–2 The Complex Number System

1. The complex plane is sectioned into four quadrants containing a real axis (the x-axis) and an imaginary axis (the y-axis). The $\pm j$ symbol is used to indicate the y-axis. Angles can be represented on the complex plane. The positive real axis represents an angle of zero degrees, the $+j$ axis represents 90°, the negative real axis represents 180°, the $-j$ axis is the 270° point, and, after a full rotation of 360°, we are back to the positive real axis.
2. A point located on the complex plane can be classified as either real, imaginary, or a combination of the two.

Section 12–3 Rectangular and Polar Forms

1. A phasor quantity is represented in *rectangular form* by the algebraic sum of the real value of the coordinate and the j value of the coordinate. An "arrow" drawn from the origin to the coordinate point in the complex plane is used to represent graphically the phasor quantity.
2. Phasor quantities can also be expressed in *polar form,* which consists of the phasor magnitude and the angular position relative to the positive x-axis.
3. Phasors can be converted between rectangular and polar forms with most scientific calculators.

Section 12–4 Mathematical Operations

1. Complex numbers must be in rectangular form in order to add or subtract them. The real parts of each complex number are added (or subtracted) to get the real part of the result. The *j* parts of each complex number are added (or subtracted) to get the *j* part of the result.

2. Multiplication or division of two complex numbers can be done in either rectangular or polar form but is most easily performed with both numbers in polar form. For multiplication, the method is to multiply the magnitudes, and add the angles algebraically. For division, divide the magnitude of the numerator by the magnitude of the denominator to get the magnitude of the quotient and subtract the denominator angle from the numerator angle to get the angle of the quotient.

3. Since sine waves can be represented by phasors, they can be described in terms of complex numbers either in rectangular or polar form.

Section 12–5 TECHnology Theory Into Practice

This section illustrates how to measure the approximate phase angle between two sine waves with the same frequency, using an oscilloscope. The time/division control is taken out of the calibrated position and varied until there is exactly one-half cycle of the sine wave displayed across the screen. Since a half cycle contains 180°, each of the ten main horizontal divisions represents 18°. After the scale has been established, the traces for both channels are superimposed and aligned along the horizontal axis to prevent any vertical offset of the waveforms. The phase angle between the two waveforms can be most easily measured from the zero crossings. The number of divisions between the two waveforms is counted and multiplied by 18° per division to read the phase angle.

Section 12–6 Phasor Analysis with PSpice and Probe

The three ac transient values introduced in this section are *delay, damping factor,* and *phase*. Delay allows you to switch different waveforms on at different times after $t = 0$. The damping factor describes how quickly the waveform dies out, or decays, after it starts up. The phase, expressed in degrees, indicates when the waveform has a positive zero-crossing relative to a reference waveform that has a positive zero at $t = 0$.

CHAPTER 13 CAPACITORS

Section 13–1 The Basic Capacitor

1. A basic capacitor is constructed of two parallel conductive plates separated by an insulating material called the dielectric. A capacitor stores energy in the form of an electric field that is established by the opposite charges on the two plates.

2. The charge per unit of voltage that a capacitor can store is its capacitance, C. The unit for capacitance is the farad, F. The basic equation for capacitance is

$$C = \frac{Q}{V}$$

where C is the capacitance in farads, Q is the charge in coulombs, and V is the voltage in volts.

3. Capacitors are rated for maximum dc voltage, temperature coefficient, and leakage current.

4. Physical parameters that determine the size (capacitance) of a capacitor are the plate area, plate separation, and the dielectric constant of the material between the plates.

Section 13–2 Types of Capacitors

1. There are several types of fixed capacitors. For non-electrolytic capacitors, the type of capacitor is generally given by the dielectric material that is used (mica, ceramic, plastic film, etc.).
2. Electrolytic capacitors are polarized (polarity *must* be observed) and offer much higher capacitance than other types, but their voltage ratings are typically lower than mica or ceramic capacitors and they have higher leakage current. Two types of electrolytic capacitors are *aluminum* and *tantalum*.
3. Variable capacitors generally have small values. Typically, they are constructed with moveable plates that mesh together.

Section 13–3 Series Capacitors

1. When capacitors are connected in series, the total capacitance is less than that of the smallest capacitor. The equation for the capacitance of series capacitors is similar to the resistance equation for parallel resistors. One way to write the equation is

$$\frac{1}{C_T} = \frac{1}{C_1} + \frac{1}{C_2} + \frac{1}{C_3} + \ldots + \frac{1}{C_n}$$

2. A series connection of charged capacitors acts as a voltage divider. The voltage across each capacitor in series is inversely proportional to its capacitance value. The largest capacitor in series will have the *smallest* voltage, and the smallest capacitor will have the *largest* voltage.

Section 13–4 Parallel Capacitors

When capacitors are connected in parallel, the effective plate area increases, and the total capacitance is the sum of the individual capacitances. That is

$$C_T = C_1 + C_2 + C_3 + \ldots + C_n$$

Section 13–5 Capacitors in DC Circuits

1. In a dc circuit, containing resistance, voltage across a capacitor *cannot* change instantaneously. Capacitors appear open to dc, and as a low impedance to a changing waveform.
2. The time constant of a series *RC* circuit is a time interval (in seconds) that equals the product of the resistance (in ohms) and the capacitance (in farads). The time constant is symbolized by τ (Greek letter tau), and the formula is:

$$\tau = RC$$

3. When a pulse is applied to an an *RC* circuit, the shape of the voltage across the capacitor is an exponential curve.

Section 13–6 Capacitors in AC Circuits

1. The instantaneous capacitor current is equal to the capacitance times the instantaneous rate of change of the voltage across the capacitor. This is written

$$i = C\frac{dv}{dt}$$

2. A sinusoidal voltage produces a sinusoidal current in a capacitive circuit. The current from the capacitor always leads the voltage by 90°.
3. Capacitive reactance is the opposition to sinusoidal current, expressed in ohms. It is applied only to a single frequency sinusoidal waveform and is expressed as

$$X_c = \frac{1}{2\pi fC}$$

where X_C is the capacitive reactance in ohms, f is the frequency of the sine wave in Hertz, and C is the capacitance in farads.
4. No power is dissipated in an ideal capacitor because current and voltage are not in phase. The product of current and voltage gives the rate that a capacitor stores or returns energy. This is called reactive power, expressed in units of volt-ampere reactive (VAR).

Section 13–7 Capacitor Applications
Applications for capacitors include power supply filters, blocking dc voltages, coupling ac signals, bypassing, timing circuits, and more.

Section 13–8 Testing Capacitors
A quick (but not comprehensive) test of a capacitor is to verify that it can be charged with an ohmmeter (through the ohmmeter's internal resistance). As the capacitor charges, the reading on the ohmmeter increases.

Section 13–9 TECHnology Theory Into Practice
The section illustrates an amplifier with coupling capacitors.

Section 13–10 Transient Analysis of *RC* Circuits with PSpice and Probe
The .AC PULSE specification is used to describe a pulse waveform. The general format of the PULSE waveform is

PULSE (<Vi> <Vp> <Td> <Tr> <Tf> <Width> <Period>)

The description indicates the initial value, pulse amplitude, delay, rise time, fall time, pulse width, and period.

CHAPTER 14 INDUCTORS

Section 14–1 The Basic Inductor
1. When a length of wire is formed into a coil, it becomes a basic inductor, symbolized by the letter *L*. Current through a coil produces an electromagnetic field. The unit of inductance is the henry, abbreviated H.
2. A changing current causes a change in the electromagnetic field, which in turn induces a voltage across the coil. The induced voltage is given by:

$$v = L\frac{di}{dt}$$

where v is the induced voltage, L is the inductance in henries, and di/dt is the rate of current change.

3. Factors that affect the amount of inductance of a coil are the core material, number of turns of wire, length, and cross-sectional area.
4. Moving a magnet through a coil of wire causes a voltage to be induced across the coil. The amount of induced voltage is directly proportional to the rate of change of the magnetic field with respect to the coil.
5. When a coil is exposed to a changing magnetic field, a voltage is induced across the coil. The induced voltage is proportional to the number of turns of wire and to the rate at which the magnetic field changes. Faraday's law expresses this as follows:

$$v_{ind} = N\frac{d\phi}{dt}$$

where v_{ind} is the induced voltage, N is the number of turns of the coil, and $d\phi/dt$ is the rate of change of the magnetic flux.

Section 14–2 Types of Inductors
Inductors are classified according to the type of material in their core and whether they are fixed or variable.

Section 14–3 Series Inductors
When inductors are connected in series, the total inductance is equal to the sum of the individual inductors. That is:

$$L_T = L_1 + L_2 + L_3 + \ldots + L_n$$

Section 14–4 Parallel Inductors
When inductors are connected in parallel, the reciprocal of the total inductance is found by adding the reciprocals of the individual inductors. That is:

$$\frac{1}{L_T} = \frac{1}{L_1} + \frac{1}{L_2} + \frac{1}{L_3} + \ldots + \frac{1}{L_n}$$

Section 14–5 Inductors in DC Circuits
1. In a dc circuit, containing resistance, current in an inductor *cannot* change instaneously. When constant direct current flows in an inductor, there is no induced voltage (except for a small drop due to the winding resistance of the coil). Ideal inductors appear as a low impedance to dc; they oppose a change in current.
2. The time constant of a series *RL* circuit is a time interval (in seconds) that equals the quotient of the inductance (in henries) divided by the resistance (in ohms). The time constant is symbolized by τ (Greek letter tau), and the formula is:

$$\tau = \frac{L}{R}$$

3. When a pulse is applied to an *RL* circuit, the shape of the current in the inductor is an exponential curve.

Section 14–6 Inductors in AC Circuits

1. The voltage induced across an inductor (Section 14–1) is determined by the rate of change of current. Hence an alternating current in an inductor induces an alternating voltage that leads the current by 90° (Alternatively, we could say that the current lags the voltage by 90°).

2. Inductive resistance is the opposition to sinusoidal current, expressed in ohms. It is applied only to a single frequency sinusoidal waveform and is expressed as

$$X_L = 2\pi fL$$

where X_L is the capacitive reactance in ohms, f is the frequency of the sine wave in Hertz, and L is the inductance in henries.

4. No power is dissipated in an ideal inductor because current and voltage are not in phase. The product of current and voltage gives the rate that an inductor stores or returns energy. This is called reactive power, expressed in units of volt-ampere reactive (VAR).

Section 14–7 Inductor Applications

Inductors have more limited applications than capacitors. They are primarily used in filter circuits (such as radio frequency blocking "chokes") and frequency selection (tuned circuits).

Section 14–8 Testing Inductors

The most common failure of an inductor is an open. This can be readily identified with a resistance check.

Section 14–9 TECHnology Theory Into Practice

This section illustrates a test of an inductor's value by measuring the time constant.

Section 14–10 Transient Analysis of *RL* Circuits with PSpice and Probe

Probe allows you to change the range of displayed waveforms, including choosing a log scale. The .TRAN directive allows you to choose the time range as well as the sample interval.

CHAPTER 15 TRANSFORMERS

Section 15–1 Mutual Inductance

When changing magnetic flux lines from one coil cut through a second coil, the two coils are said to be *magnetically coupled*. A voltage is induced in the second coil as a result of current in the first coil. This voltage is dependent on the mutual inductance, L_M, between the two coils. The mutual inductance is determined by the inductance of each coil and by the amount of coupling *(k)* between the two coils. The coefficient of coupling between the two depends on the proximity, the core material on which they are wound, and the construction and shape of the core.

Section 15–2 The Basic Transformer

1. A transformer consists of two (or more) coils that are magnetically coupled. The source voltage is applied to the primary winding (input) and the load is connected to the secondary winding (output). The windings are formed around a core that provides both the physical structure for the windings and a magnetic path.

2. Cores consist of air, ferrite, or iron. Air-core and ferrite-core transformers are used for high-frequency applications. Iron-core transformers generally are used for audio frequency (af) and power applications. These transformers consist of windings on a core constructed from laminated sheets of ferromagnetic material.

3. The turns ratio is the ratio of the number of turns in the secondary coil divided by the number of turns in the primary winding.

Section 15–3 Step-Up Transformers

1. In a step-up transformer, the output voltage is higher than the input voltage.
2. For transformers with a coefficient of coupling equal to 1, the turns ratio is equal to the voltage ratio. Thus for step-up transformers, the turns ratio is greater than 1.0.

Section 15–4 Step-Down Transformers

1. In a step-down transformer, the output voltage is lower than the input voltage.
2. For transformers with a coefficient of coupling equal to 1, the turns ratio is equal to the voltage ratio. Thus for step-down transformers, the turns ratio is less than 1.0.

Section 15–5 Loading the Secondary Winding

In an ideal transformer, the power delivered to the primary is equal to the power delivered to the load. The ideal transformer consumes no power.

Section 15–6 Reflected Load

Resistance can be thought of as the ratio of voltage to current. Since a transformer changes this ratio, it can be seen that the resistance "seen" in the primary side is not the same as the load resistance. The effective resistance seen on the primary side is called *reflected resistance,* which can be computed from the following relationship:

$$R_{pri} = \left(\frac{N_{pri}}{N_{sec}}\right)^2 R_L$$

where R_{pri} is the reflected resistance, N_{pri}/N_{sec} is the reciprocal of the turns ratio, and R_L is the load resistance.

Section 15–7 Matching the Load and Source Resistances

Certain transformers are designed to match the Thevenin source impedance with the load resistor, resulting in a maximum power transfer to the load.

Section 15–8 The Transformer as an Isolation Device

Isolation transformers electrically isolate electronic equipment from the ac power line. The isolation transformer prevents a possible shock hazard if the 110 V line is connected to the metal chassis of the equipment.

Section 15–9 Nonideal Transformer Characteristics

Real transformers have winding resistance, core losses, magnetic flux leakage, and other losses that reduce the efficiency. Most power transformers have efficiencies exceeding 95%.

Section 15–10 Other Types of Transformers

1. Many transformers have center-tapped secondaries. In addition, transformers are available with multiple windings, to allow for either different source voltages or different secondary voltages.
2. Autotransformers have a single winding that serves as both the primary and the secondary. They are frequently used to vary the ac, either above or below the input voltage.

Section 15–11 Troubleshooting Transformers

Transformer faults include open or shorted windings, or partial shorts in either winding.

Section 15–12 TECHnology Theory Into Practice
This section illustrates troubleshooting as applied to a power supply.

Section 15–13 Modeling Transformers with PSpice and Probe
PSpice uses individual inductors and a coupling coefficient to describe a transformer. The general form to describe a transformer core is

<Core name><Inductor 1><Inductor 2><Coupling>

PSpice models the inductors that make up a transformer by their inductance, not their number of turns.

CHAPTER 16 *RC* CIRCUITS

Section 16–1 Sinusoidal Response of *RC* Circuits
1. When *RC* components are connected in series to an ac source, phase shifts are introduced that depend on the amount of resistance and capacitive reactance.
2. In a capacitor, the current is always leading the voltage by 90°. In an *RC* circuit, the source current will lead the source voltage by an amount that depends on the amount of resistance and capacitive reactance.

Section 16–2 Impedance and Phase Angle of Series *RC* Circuits
1. *Impedance* is the total opposition to sinusoidal current. The total impedance is the phasor sum of the resistance and capacitive reactance. That is:

$$\mathbf{Z} = R - jX_C$$

2. *Values of R*, X_C, and **Z** can be shown graphically on an *impedance triangle*. *R* is plotted along the *x*-axis, X_C is plotted along the negative *y*-axis (shown by the *-j*) and **Z** is the vector (phasor) sum of these quantities. The phase angle can be readily found from the impedance triangle, using trigonometry.

Section 16–3 Analysis of Series *RC* Circuits
1. Ohm's law for ac circuits is **V** = **IZ.** The boldface type is used to show that both magnitude and angle are included in the calculation.
2. In any series circuit, the current is the same throughout. In a series *RC* circuit, the voltage on the resistor is in phase with the current and the voltage on the capacitor lags the current by 90°. The source voltage is the phasor sum of the voltage on the resistor and the voltage on the capacitor.
3. The graphical representation of the voltages in a series *RC* circuit has the same shape as the impedance phasor diagram. The voltage phasors can be obtained by multiplying the impedance phasors by the current in the circuit.

Section 16–4 Impedance and Phase Angle of Parallel *RC* Circuits
1. In any parallel circuit, the voltage is the same across all components. In a parallel *RC* circuit, the current in the resistor is in phase with the voltage across the resistor and the current in the capacitor leads the voltage by 90°. The source current is the phasor sum of the current in the resistor and the current in the capacitor.

2. The magnitude of the impedance in a parallel *RC* circuit can be obtained from the product over sum rule (again using phasor arithmetic). That is,

$$Z = \frac{RX_C}{\sqrt{R^2 + X_C^2}}$$

3. Conductance *(G),* Capacitive susceptance *(B_C),* and Admittance *(Y)* are the reciprocals of resistance, capacitive reactance, and impedance, respectively. Values of *G*, $B_{C,}$, and **Y** can be shown graphically on an *admittance triangle*. *G* is plotted along the *x*-axis, B_C is plotted along the positive *y*-axis and **Y** is the vector (phasor) sum of these quantities; the phase angle is found from trigonometry.

Section 16–5 Analysis of Parallel *RC* Circuits

1. In any parallel circuit, the voltage is the same across all components. In a parallel *RC* circuit, the current in the resistor is in phase with the voltage and the current in the capacitor leads the voltage by 90°. The source current is the phasor sum of the current in the resistor and the current in the capacitor.
2. The graphical representation of the currents in a parallel *RC* circuit is similar to the graphical analysis of the admittance triangle. The current phasor diagram has the same shape as the admittance phasor diagram. The current phasors can be obtained by multiplying the admittance phasors by the applied voltage.
3. Parallel *RC* circuits can be converted to an equivalent series *RC* circuit by making the impedance and phase angle, as seen by the source, the same for each circuit.

Section 16–6 Series-Parallel *RC* Circuits

Most series-parallel combination circuits can by analyzed by the method of equivalent circuits learned from series-parallel dc analysis, except phasor arithmetic must be used throughout an ac problem. Series components are combined by adding their impedances; parallel components are combined by adding their admittances. The total impedance for the circuit is found, then the circuit current. Kirchhoff's voltage law, Kirchhoff's current law, and Ohm's law can all be applied to the analysis, provided phasor arithmetic is used.

Section 16–7 Power in *RC* Circuits

For a series *RC* circuit, the power triangle can be found by multiplying the voltage phasors by the current in the circuit. For a parallel *RC* circuit, the power triangle can be found by multiplying the current phasors by the voltage across the circuit. Power in the resistor is called true power. The product of voltage and current in a capacitor is called reactive power. The product of voltage and current for the circuit is called apparent power, a more appropriate rating for generators and ac sources.

Section 16–8 Basic Applications

1. *RC* circuits form lead or lag networks, designed to shift the phase of a specific frequency by some angle. The same *RC* circuit can be used to block either low or high frequencies, depending on which component the output is taken from. The magnitude of the voltage can be found from the voltage

divider relationship (using phasor arithmetic). The equations for the ouput magnitude are:

output taken across *R*:

$$V_{out} = \left(\frac{R}{\sqrt{R^2 + X_C^2}}\right)V_{in}$$

output taken across X_C:

$$V_{out} = \left(\frac{X_C}{\sqrt{R^2 + X_C^2}}\right)V_{in}$$

2. The frequency at which the capacitive reactance is equal to the resistance is the cutoff frequency for an *RC* filter. The phase shift at cutoff is $-45°$.

Section 16–9 Troubleshooting
Typical troubles in *RC* series-parallel circuits include open and shorted components and leaky capacitors. Open components will show the source voltage across them. Shorted components will show zero volts. A leaky capacitor has an equivalent resistance in parallel with it.

Section 16–10 TECHnology Theory Into Practice
This section illustrates a capacitively coupled amplifier circuit board.

Section 16–11 PSpice Analysis of *RC* Circuits
The .AC control statement in PSpice enables you to sample data over a frequency range.

CHAPTER 17 *RL* CIRCUITS

Section 17–1 Sinusoidal Response of *RL* Circuits
1. When *RL* components are connected in series to an ac source, phase shifts are introduced that depend on the amount of resistance and inductive reactance.
2. In an inductor, the voltage is always leading the current by 90°. In an *RL* circuit, the source voltage will lead the source current by an amount that depends on the amount of resistance and inductive reactance.

Section 17–2 Impedance and Phase Angle of Series *RL* Circuits
1. *Impedance* is the total opposition to sinusoidal current. The total impedance is the phasor sum of the resistance and inductive reactance. That is:

$$\mathbf{Z} = R + jX_L$$

2. Values of R, X_L, and $\mathbf{Z}$ can be shown graphically on an *impedance triangle*. R is plotted along the *x*-axis, X_L is plotted along the positive *y*-axis (shown by the $+j$) and $\mathbf{Z}$ is the vector (phasor) sum of these quantities. The phase angle can be readily found from the impedance triangle, using trigonometry.

Section 17–3 Analysis of Series *RL* Circuits
1. As given in Chapter 16, Ohm's law for ac circuits is $\mathbf{V} = \mathbf{IZ}$. The boldface type is used to show that both magnitude and angle are included in the calculation.
2. In any series circuit, the current is the same throughout. In a series *RL* circuit, the voltage on the resistor is in phase with the current and the voltage on the inductor leads the current by 90°. The source voltage is the phasor sum of the voltage on the resistor and the voltage on the capacitor.
3. The graphical representation of the voltages in a series *RL* circuit has the same shape as the

impedance phasor diagram. The voltage phasors can be obtained by multiplying the impedance phasors by the current in the circuit.

Section 17–4 Impedance and Phase Angle of Parallel *RL* Circuits

1. In any parallel circuit, the voltage is the same across all components. In a parallel *RL* circuit, the current in the resistor is in phase with the voltage across the resistor and the current in the capacitor leads the voltage by 90°. The source current is the phasor sum of the current in the resistor and the current in the inductor.

2. The magnitude of the impedance in a parallel *RL* circuit can be obtained from the product over sum rule—the sum is a vector sum expressed as the square root of the squares. That is

$$\mathbf{Z} = \frac{RX_L}{\sqrt{R^2 + X_L^2}}$$

3. Conductance *(G)*, Inductive susceptance (B_L), and Admittance *(Y)* are the reciprocals of resistance, inductive reactance, and impedance, respectively. Values of *G*, B_{L}, and **Y** can be shown graphically on an *admittance triangle*. *G* is plotted along the *x*-axis, B_{L} is plotted along the negative *y*-axis and **Y** is the vector (phasor) sum of these quantities; the phase angle is found from trigonometry.

Section 17–5 Analysis of Parallel *RL* Circuits

1. In any parallel circuit, the voltage is the same across all components. In a parallel *RL* circuit, the current in the resistor is in phase with the voltage and the current in the inductor lags the voltage by 90°. The source current is the phasor sum of the current in the resistor and the current in the inductor.

2. The graphical representation of the currents in a parallel *RL* circuit is similar to the graphical analysis of the admittance triangle. The current phasor diagram has the same shape as the admittance phasor diagram. The current phasors can be obtained by multiplying the admittance phasors by the applied voltage.

3. Parallel *RL* circuits can be converted to an equivalent series *RL* circuit by making the impedance and phase angle, as seen by the source, the same for each circuit.

Section 17–6 Series-Parallel *RL* Circuits

As in the case of combination *RC* circuits, most combination *RL* circuits can be analyzed by the method of equivalent circuits with care to use phasor quantities throughout the analysis. Kirchhoff's voltage law, Kirchhoff's current law, and Ohm's law can all be applied to the analysis, as mentioned previously.

Section 17–7 Power in *RL* Circuits

For a series *RL* circuit, the power triangle can be found by multiplying the voltage phasors by the current in the circuit. For a parallel *RL* circuit, the power triangle can be found by multiplying the current phasors by the voltage across the circuit. Power in the resistor is called true power. The product of voltage and current in a capacitor is called reactive power. The product of voltage and current for the circuit is called apparent power.

Section 17–8 Basic Applications

1. As in the case of *RC* circuits, *RL* circuits can be designed to shift the phase of a specific frequency by some specified angle. The same basic *RL* circuit can be used to block either low or high

frequencies. The magnitude of the voltage can be found from the voltage divider relationship (using phasor arithmetic). The equations for the output magnitude are:

output taken across R:

$$V_{out} = \left(\frac{R}{\sqrt{R^2 + X_L^2}}\right)V_{in}$$

output taken across X_C:

$$V_{out} = \left(\frac{X_L}{\sqrt{R^2 + X_L^2}}\right)V_{in}$$

2. The frequency at which the inductive reactance is equal to the resistance is the cutoff frequency for an RL filter, and it causes a phase shift of 45°.

Section 17–9 Troubleshooting
Troubles in ac series-parallel RL circuits include open and shorted components; however, a shorted inductor is an uncommon fault. As is the case with any open, the source voltage will be found across the open component.

Section 17–10 TECHnology Theory Into Practice
This section illustrates an RL filter module.

Section 17–11 PSpice Analysis of RL Circuits
An example of an RL circuit using the .AC control statement to sample data over a linear frequency range is given.

CHAPTER 18 *RLC* CIRCUITS AND RESONANCE

Section 18–1 Impedance of Series RLC Circuits
1. A series RLC circuit contains resistance, inductance, and capacitance. The capacitive reactance and inductive reactance tend to offset each other. Depending on which impedance is *larger,* the circuit can appear inductive or capacitive.
2. The total impedance of a series RLC circuit is the phasor sum of the reactive and resistive elements. That is:

$$\mathbf{Z} = R + jX_L - jX_C \quad \text{(rectangular form)}$$
$$\mathbf{Z} = \sqrt{R^2 + (X_L - X_C)^2}\ \angle\tan^{-1}\left(\frac{X_T}{R}\right) \quad \text{(polar form)}$$

Section 18–2 Analysis of Series RLC Circuits
1. Recall that the capacitive and inductive reactance depend on frequency. As the frequency is raised, the capacitive reactance decreases and the inductive reactance increases. *Series resonance* occurs when they are equal.
2. In a series RLC circuit, the voltage across the capacitor and inductor are 180° out of phase with each other and tend to reduce the voltage across both components. Thus, the source voltage can be lower than the voltage across the capacitor or the inductor.
3. To solve for the voltages across the components, first find the total impedance of the circuit (see section 18–1). Then apply Ohm's law to find the current throughout the circuit. Find the voltage across each component by Ohm's law.

Section 18–3 Series Resonance

1. The frequency at which $X_C = X_L$ is called the *resonant frequency,* f_r, which is true for both series and parallel resonance (ignoring the effect of coil resistance). In a series resonant circuit, the capacitive and inductive reactances cancel each other and the circuit appears to be purely resistive (phase angle between source current and source voltage = 0°). The total impedance is equal to the resistance. Note that the voltage across the capacitor tends to cancel the voltage across the inductor.

2. Ignoring the coil resistance, the resonant frequency is computed from the equation

$$f_r = \frac{1}{2\pi\sqrt{LC}}$$

3. At series resonance, the impedance of the circuit is a minimum and the current is maximum. Above or below the resonant frequency, the impedance increases and the current in the circuit decreases.

Section 18–4 Impedance of Parallel *RLC* Circuits

1. In a parallel *RLC* circuit the circuit can appear to be either capacitive or inductive depending on which impedance is *smaller,* since a smaller impedance in parallel will have a greater current in it.

2. It is convenient to consider a parallel *RLC* circuit as being composed of parallel branches composed of conductance (reciprocal resistance), capacitive susceptance, and inductive susceptance (reciprocal reactance). The total admittance is the phasor sum of the conductance and susceptances. That is:

$$\mathbf{Y} = G + jB_C - jB_L \quad \text{(rectangular form)}$$

Section 18–5 Analysis of Parallel and Series-Parallel *RLC* Circuits

1. For purely parallel circuits, the voltage is the same across all components, so analysis is simplified by finding the current in each component using Ohm's law. Then add the branch currents (using phasors) to find the total source current. In polar form, the sum of the branch currents is given by:

$$\mathbf{I}_T = \sqrt{I_R^2 + (I_C - I_L)^2}\ \angle\tan^{-1}\left(\frac{I_{CL}}{I_R}\right) \quad \text{(polar form)}$$

where I_{CL} is the current in *both* the inductor and the capacitor.

2. There are many variations of series-parallel combination circuits; hence the method for analysis is dependent on the circuit and what is known. Generally, it is possible to find the total impedance by combining components into equivalent circuits. The total current can then be found, and the analysis proceeds by applying Kirchhoff's voltage and current laws and Ohm's law as illustrated in previous chapters. Care must be taken to use phasor quantities throughout the analysis.

Section 18–6 Parallel Resonance

1. As in the case of series resonance, the frequency at which $X_c = X_L$ is the *resonant frequency,* f_r (again ignoring the effect of coil resistance). In a parallel resonant circuit, the capacitive and inductive

susceptances tend to cancel. Note that the current in the capacitor tends to cancel the current in the inductor.
2. Ignoring coil resistance, the resonant frequency is computed from the equation

$$f_r = \frac{1}{2\pi\sqrt{LC}}$$

3. At parallel resonance, the impedance of the circuit is a maximum and the current is a minimum. Above or below the resonant frequency, the impedance decreases and the source current increases.
4. The Q of a coil is the ratio of the reactance of the coil to its resistance. The effect is more pronounced in low Q circuits and affects the response as well as the resonant frequency. A parallel resistor will also tend to lower the Q of the circuit.

Section 18–7 Bandwidth of Resonant Circuits
1. The bandwidth, *BW,* of a resonant circuit is a function of the Q of the circuit. Bandwidth is the difference between the upper and lower critical frequencies. The upper and lower critical frequencies are the frequencies at which power from the source is one half that of the maximum.
2. A circuit which has a high Q has a smaller bandwidth and is said to have a high selectivity. Selectivity is a measure of how well a circuit tunes one frequency to the exclusion of other frequencies.

Section 18–8 System Applications
Applications for resonant circuits include tuned amplifiers, such as found in a superheterodyne receiver. The resonant circuit selects the frequency of the station that is received. In the application illustrated, a varactor diode is used for changing the capacitor in the resonant circuit.

Section 18–9 TECHnology Theory Into Practice
This section illustrates the superheterodyne receiver using a varactor for tuning.

Section 18–10 PSpice Analysis of *RLC* Circuits
The bandwidth of a resonant circuit can be observed in PSpice by viewing the voltages across the components as a function of frequency by using Probe. At resonance, the phase angle between source voltage and current is zero; this can be observed with the IP(VS) statement.

CHAPTER 19 BASIC FILTERS

Section 19–1 Low-Pass Filters
1. A basic low-pass filter consists of a series *RC* circuit with the output taken across the capacitor (or an *RL* series circuit with the output taken across the resistor).
2. The range of low frequencies passed by a low-pass filter within a specified limit is called the pass band of the filter. The point considered to be the upper end of the band pass is at the *critical frequency,* f_c. At the critical frequency, $X_C = R$ (for basic *RC* filters) or $X_L = R$ (for basic *RL* filters). The critical frequency is the frequency at which the filter's output voltage is 70.7% of the maximum.
3. The *decibel* is a logarithmic measurement of the ratio of one power to another or one voltage to another as expressed in the equations:

$$\text{dB} = 10 \log\left(\frac{P_{\text{out}}}{P_{\text{in}}}\right) \qquad \text{dB} = 20 \log\left(\frac{V_{\text{out}}}{V_{\text{in}}}\right)$$

4. For each tenfold *(decade)* increase in frequency above f_c, there is a 20 dB reduction in the output voltage for a basic *RC* or *RL* filter. The response plotted on semilog paper, with frequency on the log scale, is called a *Bode plot*.
5. Both low-pass and high-pass filters shift the phase of the signal. At the critical frequency, the phase shift is 45°. Low-pass filters are basic lag networks and shift the phase in the negative direction.

Section 19–2 High-Pass Filters

1. A basic high-pass filter consists of a series *RC* circuit with the output taken across the resistor (or an *RL* series circuit with the output taken across the inductor).
2. As in the case of low-pass filters, the point at which $X = R$ is at the *critical frequency*, f_c. Below the critical frequency the response of basic filters drops 20 dB per decade.
3. High-pass filters are basic lead networks and shift the phase in the positive direction.

Section 19–3 Band-Pass Filters

1. The bandwidth, *BW*, of a band-pass filter is the range of frequencies between the upper and lower critical frequencies.
2. A band-pass filter can be constructed from either a series resonant circuit or a parallel resonant circuit or from a series combination of low-pass and high-pass filters.
3. A series resonant bandpass filter consists of a resistor, capacitor, and inductor. The circuit has minimum impedance and maximum current at the resonant frequency, f_c. The output from a series resonant bandpass filter is taken across the resistor.
4. A parallel resonant bandpass filter consists of a series resistor and a parallel inductor and capacitor *(tank circuit)*. The circuit has maximum impedance and minimum current at the resonant frequency, f_c. The output from a parallel resonant bandpass filter is taken across the *LC* tank circuit.
5. The bandwidth of resonant filters is determined by the circuit Q. A higher value of circuit Q results in a smaller bandwidth and a lower value of Q causes a larger bandwidth, as given in the following equation:

$$BW = \frac{f_0}{Q}$$

where *BW* is the bandwidth and f_0 is the resonant frequency.

Section 19–4 Band-Stop Filters

1. A band-stop filter allows all frequencies to pass except those lying within a certain stop band defined by the upper and lower critical frequencies. Band-stop filters can be constructed from either series or parallel resonant circuits or from a parallel combination of low-pass and high-pass filters.
2. The series band-stop filter is equivalent to the series bandpass circuit, except the output of the band-stop filter is taken across the inductor and capacitor (the *LC* resonant circuit).
3. The parallel band-stop filter is the same as the parallel bandpass circuit, except the output of the band-stop filter is taken across the resistor.

Section 19–5 TECHnology Theory Into Practice

This section illustrates testing basic filters.

Section 19–6 Filter Analysis with PSpice and Probe

This section illustrates how two separate filter circuits can be analyzed by PSpice in the same circuit file. Probe will draw output voltage traces for both filters, making comparisons easy. Probe allows you to do separate traces using the "Plot-control" option.

CHAPTER 20 CIRCUIT THEOREMS IN AC ANALYSIS

Section 20–1 The Superposition Theorem

1. The superposition theorem can be stated as follows:

The current in any given branch of a multiple-source circuit can be found by determining the currents in that particular branch produced by each source acting alone, with all other sources replaced by their internal impedances. The total current in the given branch is the phasor sum of the individual source currents in that branch.

2. As in dc circuit analysis, the superposition theorem is applied by solving for each source, one at a time, when the other sources are replaced with their internal series impedance. Currents due to each source are computed in the branch or branches of interest, and the individual currents are added as phasor quantities.

Section 20–2 Thevenin's Theorem

1. A Thevenin circuit can be applied to ac circuits and can replace any linear, two-terminal circuit with an equivalent ac voltage source (designated $\mathbf{V}_{th}$) in series with an equivalent impedance (designated $\mathbf{Z}_{th}$). The equivalent impedance can be any combination of resistance and reactance. For *any load* connected to the Thevenin circuit, the current and voltage in the load is identical to the original circuit.
2. $\mathbf{V}_{th}$ is defined as the open circuit voltage between two specified terminals in an ac circuit.
3. $\mathbf{Z}_{th}$ is defined as the total impedance appearing between two specified terminals in an ac circuit with all sources replaced with their internal impedance.

Section 20–3 Norton's Theorem

Norton's theorem is similar to Thevenin's theorem, but it reduces a complex circuit to an equivalent current source (designated $\mathbf{I}_n$) with an equivalent *parallel* impedance (designated $\mathbf{Z}_n$). The current source for Norton's circuit is equal to the amount that would flow if a short were placed across the output terminals. The Norton impedance is exactly the same as the Thevenin impedance.

Section 20–4 Millman's Theorem

1. Millman's theorem permits any number of parallel branches consisting of voltage sources and impedances to be reduced to a single equivalent voltage source and equivalent series impedance.
2. Notice that with Millman's theorem, you could replace each parallel branch with a Norton circuit, then combine the current sources by algebraic addition and combine the parallel impedances in the usual way (by adding the reciprocals and taking the reciprocal of the result). The result is a "combined" Norton circuit that can be converted to the series Thevenin equivalent.

Section 20–5 Maximum Power Transfer Theorem

Maximum power transfer occurs when the load is made the complex conjugate of the source (assuming the source impedance cannot be changed). The complex conjugate has the same value of resistance and the same *magnitude* of the reactive term but the opposite direction. This means that the reactive terms tend to cancel (equivalent to a series resonant circuit).

Section 20–6 TECHnology Theory Into Practice

This section illustrates a filter problem; maximum power transfer is computed.

Section 20–7 AC Equivalent Circuits Using PSpice and Probe

1. In dc analysis, the Thevenin equivalent voltage (as seen by R_L) is determined in PSpice by using a very large resistance to simulate an open. The Norton equivalent current is found by using a very

small resistance to simulate a short. The Thevenin voltage is then divided by the Norton current to get the Thevenin (or Norton) equivalent resistance.

2. In ac analysis, the equivalent values must be determined at some specified frequency, and both the magnitude and phase angle of the open circuit voltage and short circuit current must be determined.

CHAPTER 21 PULSE RESPONSE OF REACTIVE CIRCUITS

Section 21–1 The *RC* Integrator

A basic *integrator* is a series *RC* circuit with the output taken across the capacitor. It is a low-pass filter circuit that is analyzed in this chapter in terms of its pulse response. The voltage on a capacitor follows an exponential curve when the pulse is applied to the input.

Section 21–2 Single-Pulse Response of *RC* Integrators

1. When a positive pulse is applied to the input of an *RC* integrator, the capacitor begins charging from its low level toward its high level. When the pulse goes from its high level back to its low level, the capacitor discharges back through the source.

2. The shape of the capacitor waveform depends on the pulse width and the *RC* time constant (τ). When τ is long compared to the pulse width ($t_w > 5\tau$), the capacitor charges and discharges fully; otherwise the capacitor does not fully charge.

Section 21–3 Repetitive-Pulse Response of *RC* Integrators

1. When a periodic pulse waveform is applied to an *RC* integrator, the capacitor will charge and discharge in response to the pulse input.

2. When the pulse width is long compared to the *RC* time constant, the output will be a series of pulses with exponential rising and falling edges. When the pulse width is short compared to the *RC* time constant, the output will approach a constant average value in approximately 5 time constants.

Section 21–4 Single-Pulse Response of *RC* Differentiators

1. A basic *differentiator* is a series *RC* circuit with the output taken across the resistor. It is a high-pass filter circuit that will be analyzed in this chapter in terms of its pulse response. The voltage on a resistor is determined by the charging current for the capacitor, and depends on the *RC* time constant.

2. When a positive pulse is applied to the input of an *RC* differentiator, capacitor charging current flows immediately and decays as the capacitor charges. The voltage across the resistor is proportional to this charging current. When the pulse goes from its high level back to its low level, the capacitor discharge current flows in the opposite direction. It starts high and decays as the capacitor discharges.

Section 21–5 Repetitive-Pulse Response of *RC* Differentiators

1. When a square wave is applied to the *RC* differentiator, the output shape depends on the *RC* time constant. When the time constant is short compared to the pulse width, the output is a series of positive and negative "spikes."

2. When the time constant is increased, the positively and negatively sloping portions of the output waveform become flatter. For a very long time constant, the output approaches the shape of the input, but with an average value of zero (no dc component in the output).

Section 21–6 Pulse Response of *RL* Integrators

A basic integrator can be constructed from a series *RL* circuit with the output taken across the resistor. The response of a basic *RL* integrator to a single pulse is essentially identical to the response of the *RC* integrator.

Section 21–7 Pulse Response of *RL* Differentiators
A basic differentiator can be constructed from a series *RL* circuit with the output taken across the inductor. The response of a basic *RL* differentiator to a single pulse is essentially identical to the response of the *RC* differentiator.

Section 21–8 Relationship of Time (Pulse) Response to Frequency Response
1. Integrators (both *RC* and *RL*) act as low-pass filters. Differentiators act as high-pass filters.
2. The rise time of a pulse is related to the highest (meaningful) frequency component. The relationship is expressed by the (approximate) formula:

$$t_r = \frac{0.35}{f_h}$$

where t_r = rise time of the pulse and f_h = highest frequency component.

Section 21–9 Troubleshooting
As with any electronic circuits, potential problems in a basic *RC* circuit include open and shorted components.

Section 21–10 TECHnology Theory Into Practice
This section illustrates how a variable time delay circuit can be constructed from a basic integrator.

Section 21–11 Applying PSpice and Probe to Pulsed Circuits
A repetitive pulsed source can be specified in detail in PSpice as an independent voltage source. The pulse parameters specified include the initial and final voltage, delay, rise time, fall time, pulse width, and period. Voltage parameters have units of volts; time parameters have units of seconds.

CHAPTER 22 POLYPHASE SYSTEMS IN POWER APPLICATIONS

Section 22–1 Basic Polyphase Machines
1. A basic two-pole, three-phase generator has three separate fixed **field windings** placed at 120° intervals around a rotating electromagnet, called the **rotor**. This configuration induces three sinusoidal voltages separated by phase angles of 120° in the field windings as the magnetic field rotates.
2. The electromagnet is created by passing direct current (I_F) through a winding around the rotor. The direct current is applied through a brush and slip ring assembly.
3. The three-phase induction motor consists of a stator with field windings and a rotor assembly constructed as a cylindrical frame of metal bars (called a **squirrel cage**). Three-phase voltages applied to the stator windings cause a rotating magnetic field to induce current in the metal bars of the rotor. The interaction of the induced currents and the magnetic field produces forces that cause the rotor to rotate.

Section 22–2 Polyphase Generators in Power Applications
1. The size of copper conductors for load current can be reduced when a polyphase generator is used when compared to a single-phase generator. When the current in each winding of a polyphase generator is equal in magnitude, a **balanced load** condition is said to exist.
2. A second advantage of a polyphase system is that the power in the load is constant.
3. A third advantage of a polyphase system is that the force that rotates the shaft of a motor is constant, producing constant shaft rotation.

Section 22–3 Three-Phase Generators

1. The Y-connected system has the following attributes:
   (a) The system can be either a three-wire or, when the neutral is used, a four-wire system connected to load. With a balanced load, the neutral current is zero. With an unbalanced load, a neutral wire is required to provide path for return current. Two voltage magnitudes are available at the terminals in the four-wire system: the phase voltage and the line voltage.
   (b) Phase voltages (V_θ), are measured across the generator windings with the neutral as reference; line voltages are measured between two lines (*a* to *b*, *a* to *c*, *b* to *c*) and are larger. Line voltage is equal to the phase voltage times $\sqrt{3}$ (1.73). Phase currents are equal to line currents.
   (c) The phasor diagram for the phase voltages shows three equal voltages separated by 120°.
   (d) The phasor diagram for the line voltages shows three equal voltages separated by 120° and larger than the phase voltages. The line voltage phasors are rotated 30° from the phase voltage phasors.
2. The delta (Δ) connected generator has the following attributes:

(a) The system is set up only as a three-wire system; only one voltage magnitude is available at the output. The magnitudes of the line voltages and phase voltages are equal, but the line currents do not equal the phase currents.
(b) The magnitude of each line current is equal to $\sqrt{3}$ times the magnitude of the phase current and that there is a phase angle of 30° between each line current and the nearest phase current.
(c) The phasor diagram for the phase currents shows three equal currents separated by 120°.
(d) The phasor diagram for the line currents shows three equal currents separated by 120° and larger than the phase currents. The line current phasors are rotated 30° from the phase current phasors.

Section 22–4 Three-Phase Source/Load Analysis

1. Three-phase systems can be connected in any combination of a Δ or Y for the source and the load. A Y-connected source driving a Y-connected load (Y-Y) can be connected to a balanced three-phase load or three independent single-phase loads. The Y-Y connection with unbalanced loads produces a non-zero neutral current.
2. The Y-Δ configuration produces the full line voltage across the load for each phase.
3. In a Δ-Y configuration, line voltages are equal to the corresponding phase voltages of the source and each phase voltage equals the difference of the corresponding load voltages.
4. In a Δ-Δ configuration, line, load, and source phase voltages are all equal for a given phase. With a balanced load, all of these voltages are equal.

Section 22–5 Three-Phase Power

1. Each phase of a balanced three-phase load has an equal amount of power; therefore, the total true load power is three times the power delivered by each phase.
2. Power is measured with wattmeters; a wattmeter senses both current and voltage in the load. Two methods for measuring power in a three-phase system are the two-wattmeter method and the three-wattmeter method.

Section 22–6 Polyphase Circuit Modeling with PSpice

PSpice can be used to model three phase circuits by assigning three separate sources, with a 120° phase shift assigned between the sources.

Appendix B
List of Materials for the Experiments

Resistors:
All resistors can be ¼ W. *Quantity:* one of each except where noted:

- 47 Ω (two)
- 100 Ω
- 150 Ω
- 270 Ω
- 330 Ω (two)
- 470 Ω
- 560 Ω
- 680 Ω
- 820 Ω
- 1.0 kΩ (two)
- 1.2 kΩ
- 1.5 kΩ
- 1.6 kΩ
- 1.8 kΩ
- 2.2 kΩ
- 2.7 kΩ
- 3.3 kΩ
- 3.6 kΩ
- 4.7 kΩ
- 5.6 kΩ
- 6.8 kΩ
- 9.1 kΩ
- 10 kΩ (two)
- 33 kΩ
- 47 kΩ
- 100 kΩ

Variable resistors:
1.0 kΩ, 5.0 kΩ, 10 kΩ

Capacitors:
All capacitors are 35 WV. *Quantity:* one of each except where noted:

- 1000 pF (two)
- 2000 pF
- 4700 pF
- 0.01 μF (two)
- 0.033 μF
- 0.047 μF
- 0.1 μF
- 1.0 μF
- 47 μF
- 100 μF

Inductors:

- One 7 H Triad C-8X (or equivalent) (second 7 H inductor optional)
- Two 100 mH

Transformers: (one of each)

- 12.6 V center-tapped: Triad F-70X (or equivalent)
- Small-impedance matching transformer (approximately 600 Ω to 8 Ω)

Transistors:

- One MPF102 *n*-channel JFET
- See optional material for 2N3904

Miscellaneous: (one of each)

- LEDs (one red, one green)
- Meter: dc ammeter 0–10 mA
- Metric ruler
- Neon bulb (NE-2 or equivalent)
- Relay: DPDT, 6 V dc or 12 V dc coil
- Small speaker (4 or 8 Ω)
- Switch: SPST

Optional Material: Materials needed for For Further Investigations and Application Problems: (one of each)

- Analog ohmmeter
- CdS photocell: Radio Shack 276-116 (or equivalent)
- Compass and protractor
- Decade resistance box
- I.F. transformer: J.W. Miller 8812 (or equivalent)
- Inductor (any value—from 1–100 mH for use as unknown)
- Meter calibrator
- 7414 Schmitt trigger
- Sweep generator for audio frequencies (to 10 kHz)
- Transistor—2N3904 npn (or equivalent)
- Wheatstone bridge
- Zener diode: 5 V